ESSENTIALS OF
FOREST SCIENCES

ESSENTIALS OF
FOREST SCIENCES

Dr. M.P. Singh, FMA
Head of the Department Forest Sciences
Birsa Agricultural University, Ranchi - 834006 (Jharkhand)

Dr. J.L. Srivastava, IFS
Principal Chief Conservator of Forests (Retd.)
Government of Jharkhand, Ranchi - 834002

Dr. R.B. Sah, Ph.D.
Department of Silviculture
Birsa Agricultural University, Ranchi - 834006

SATISH SERIAL PUBLISHING HOUSE
SSPH
403, Express Tower, Commercial Complex, Azadpur, Delhi - 110033 (INDIA)
Phone : 011-27672852 Fax : 91-11-27672046
E-mail : info@satishserial.com, hkjain1975@yahoo.com
Website : www.satishserial.com

Published by :

SATISH SERIAL PUBLISHING HOUSE

403, Express Tower, Commercial Complex, Azadpur, Delhi - 110033 (INDIA)
Phone : 011-27672852 Fax : 91-11-27672046
E-mail : info@satishserial.com, hkjain1975@yahoo.com

ISBN 81-89304-30-5

Printed at
SALASAR IMAGING SYSTEMS, DELHI-35

प्रो. वी. एल. चोपड़ा
PROF. V. L. CHOPRA

सदस्य योजना आयोग
योजना भवन, नई दिल्ली-110 001
PLANNING COMMISSION
YOJANA BHAWAN
NEW DELHI - 110 001

FOREWORD

The Forestry research in India has now history of mare than a century. Forests are vanishing at the rate of 16.8 million hectares a year. Deforestation deprives poor people, of fuel wood and produces its own environmental refugees. Deforestation in the hills causes avalanches and flooding in the plains during monsoon. When the Himalayas were covered with forests, Bangladesh suffered a major flood about twice in a century now the frequency is one in four years. Deforestation is also responsible for addition of 33 per cent of Co2 to the atmosphere (5.5 billion tons each year) which contributes to global warming. Forests are deemed to be "global sinks" for CO2 and loss of forest cover raises Co2 level in the atmosphere. Forests prevent soil erosion and provide one of nature's principal means of managing water is a high priority "area.

The deficiency will reduce our ability to maintain or enhance forest productivity, recreation and conservation as well as our ability to ameliorate or adapt to changes in the global environment.

This will help the forestry research to increase the breadth of research areas covered and the depth to which they are investigated. To achieve this Dr. M. P. Singh, Dr. J. L. Srivastava and Dr. R.B. Sah have made an attempt to put together several important research guidelines.

I hope, this book will serve to further the cause of Forestry in general and Forestry Research in particular.

New Delhi
Dated 10.1.2006

(V.L. Chopra)

PREFACE

Outstanding work has been done in forestry research in India in the past now a history of more than a century. However, current knowledge and pattern of research will not result sufficiently in accurate predictions of the consequences of potentially harmful influences in Managing the forest that lack a sound basis in biological knowledge. The deficiency will reduce our ability to maintain or enhance forest productivity, recreation and conservation as well as our ability to ameliorate or adapt to changes in the global environment. A new research paradigm will have to be adopted an environmental paradigm. This will require forestry research to increase the breadth of research areas covered and the depth to which they are investigated.

Many informations on the development of research analysis in Forest sciences are scattered into the literature and number of methods are available for research analysis of plants, water and soils. But, the analyst is confronted by some problems. The first problem is, to select a procedure, which will properly evaluate the related parameter. The second problem is, to discover appropriate contents, which are present in the experimental samples one is attempting to analyse. The third problem is, to find the most specific methods, appropriate to the required analysis.

This manual will serve to further the cause of Forestry in general and Forestry Research in particular.

Ranchi
Dated 15.1.2006

M.P.Singh
J.L. Srivastava
R.B.Sah

CONTENTS

1

Studies of Forest Products

GENERAL INTRODUCTION

Traditionally forest have been used for many products, including timber, Fuel and fodder. The biomas determining of forests is a useful way of providing estimates on the quantity of these components. The volume estimations have assessed the quantity of saw timber. Typically but this ignores the other useful components such as smaller size wood for fuel use. Furthermore, very few to none assessments have been made of the quantity of wood present in forests that appear to have no potential far saw timber production. The total above ground biomass, of forests, which is defined as biomass density when assessed, may be expressed as dry weight per unit area and is a useful way of quantifying the amount of resource available for all traditional uses. It either gives quantity of total biomass directly or the quantity of each components (*e.g.*leaves, branches and bole) because their biomass tends to vary systematically with the total biomass. The biomass of each forest component depends on the forest type, such as natural or planted forests and closed or open forest. As for example, leaves for fodder are about 4-8% and merchantable bole is about 70% of the total above ground biomass in closed forests.

Biomass quantity in a forest is the result of the difference between production through photosynthesis and consumption by respiration and harvest processes. It is a useful measure for assessing changes in forest structure. Changes Forest biomass density are brought about by natural succession, human activities such as silviculture, harvesting, and degradation; and natural impacts by wildfire and climate change. Biomass density is also a useful variable for comparing structural and functional attributes of forest ecosystems across a wide range of environmental conditions

Biomass of Global changes is very relevant issue related to. As for example, the role of tropical forests in global bio-geochemical cycles, especially the carbon cycle and its relation to the greenhouse effect, has kindled interest in estimating the biomass density of tropical forests. The biomass of forests provides estimates of the carbon pools in forest vegetation because about 50% of it is composed of carbon. Consequently, biomass represents the potential amount of carbon that can be added to the atmosphere as carbon dioxide when the forest is cleared or burnt. Attempts to estimate the biomass density of tropical forests have been made by the scientific community for use in models that assess the contribution of tropical deforestation and biomass burning in the increase of atmospheric carbon dioxide and other trace gases (Brocon et al. 1989, Brown and Iverson, 1992, FAO, 1993).

In climate change Global interests led to the establishment of the UN Framework Convention on Climate Change (UNFCCC) at the 1992 UN conference on Environment and Development (UNCED). Over 130 nations have ratified this convention, which means that these nations need to make national greenhouse gas emission inventories. Changes in the cover, use, and management of forests produce sources and sinks of carbon dioxide to and from the biosphere. To estimate the magnitude of these sources and sinks, requires reliable estimates of the biomass density of the forests undergoing change.

Estimate of biomass also provides the means for calculating the amount of carbon dioxide that can be removed from the atmosphere by rejuvenating forests or by plantations, because they establish the rates of biomass production and the upper bounds for carbon sequestering.

Issue receiving is more attention of late as countries look to forests as a means of mitigating greenhouse gas emission, particularly carbon dioxide, a major greenhouse gas and the one fixed during photosynthesis. Practices such as sustainable forest management slow down deforestation and low impact logging decrease carbon dioxide emissions and conserve it. Other practices such as plantation establishment or other tree planting programs on previously non-forested land sequester carbon dioxide (Brown et al.1996). Furthermore, biomass density estimates of forests are extremely relevant for studying other global bio-geochemical cycles, such as nitrogen, because the amount of other nutrient elements in forests is also related to the quantity of biomass present.

The forest biomass another issue emerged since the 1980s. due to loss of forest area, forest degradation, resulting in biomass density loss, is known to be occurring (Brown et al., 1994, FAO 1993, 1995). Much of this biomass degradation appears to be unrecorded; this is in addition to that accounted for by sanctioned harvesting. Art example is the illegal logging practices occurring in many forests of tropical Asia (Callister 1992). Clearly, this process of biomass density reduction has implications on the global carbon cycle, other bio-geochemical cycles, and biodiversity. Biomass degradation depend on factors mostly related to social, economic and political conditions.

About the Biomass

So many issues raised above related to forest biomass. To estimate all of forest components, including the above-ground and below-ground living mass of trees, shrubs, palms, saplings, other under storey components that change with time and with natural and-human disturbances. However for practical reasons, this methodology is restricted to the following component of biomass density: the total amount of live above-ground organic matter present in trees including leaves, twigs, branches, main bole, and bark. The biomass of these components generally accounts for the greatest fraction of total living biomass in a forest and does not pose too many logistic problems in its estimation. Consequently, biomass is defined as the total amount of above-ground living organic matter in trees expressed as oven-dry tons per unit area (tree, hectare, region or country). It is referred to as biomass density when expressed as mass per unit area, e.g., tons per hectare. The total biomass for a region or country is obtained from the product of biomass density and the corresponding area of forests.

Most of the forests or tree formations, biomass density estimates will be based only on the biomass in trees with diameters greater than or equal to 10 cm, which is the usual minimum diameter measured in most inventories of closed forests. However, for forests or trees of smaller stature, such as those in the arid or montane zones, degraded forests, or secondary forests, lower minimum diameters should be chosen.

The Woody biomass for many traditional uses does not fully originate from forests; significant quantities are obtained from non-forest lands such as small woodlots, windbreaks, other tree line formations, home gardens, etc.

Methodology Adopted

The vegetation woody

A desired information is the above -ground woody biomass, but since it is seldom practical to measure this directly, an indirect approach using the stem basal area (m^2) and height (m) of a sample of woody plants is used. In order to extrapolate the sample to a per hectare basis, the total basal area per species per hectare (m^2/ha) and the plant density (plants/ha) are also required

The Species

THE Species should be identified by their full scientific name (Genus, species, Author). Unidentified species should be given a code or local name, and a reference to a specimen in a herbarium.

A natural communities with a large number of species, it is not necessary to record the height and basal area of minor species individually. All the species contributing less than 10% of the total biomass can be combined into one category.

An in the extremely diverse communities with no clear biomass dominance, where no species contributes more than 10% to the total biomass, it may be necessary to define the 'species' on a functional basis; for instance emergent, canopy trees, sub-canopy trees and under storied shrubs, recording which species were placed in which category.

However, if the trees within the study site can be individually counted and measured. Usually this is not the case, and one or more sub samples each including about 100 trees is taken.

In this a square quadrat of known area can be laid out with the aid of a tape measure, a compass and string, but often a circular quadrat (all the trees within a given distance of a center peg) is easier to work with.

Than a long thin quadrat (belt transect) is often an efficient method in dense vegetation. Lay out a measuring tape (50 m) in a straight line, and record all the trees within a given distance (1-5 m, measured with a handheld rod) on either side of the tape. Both the length and width of transect can be varied to give a total sample of about 100 plants; different widths can be used for common shrubs and sparse trees within the same sample. Individuals (or multi stemmed clumps) more than half within the plot boundary should be counted in. To permit biomass estimates based on individual trees at a later stage, data should be recorded as in the following example:

5m x 50 m Plots

Species	**Height (m)**	**Diameter (cm)**	**Stem area (cm^2)**
Burkea african	3.2	15	176
Ochoa pulshra	1.5	5	19
Sclerocarya birrea	6	30	707

The Stem density: (i) Plotless methods

Plots Laying is time consuming and can lead to error due to subjective placement of the plots. A variety of plotless method is available, such as the Point-Center-Quarter method (PCQ; Cottam and Curtis, 1956). These methods use the distance between random points and the nearest plant to calculate an average inter-plant distance (d_x). The number of plants per hectare (N) is then given by

$$N = 10{,}000/dx^2$$

where $\dot{d}_x$ is in meters.

In an about 100 distance measurements are required, and at the same time the species, stem diameters and heights of this random sample of plants are recorded respecteively.

The PCQ method is not applicable to vegetation, which is regularly spaced (such as plantations) or highly clumped; various modifications, such as the two-distance methods of Diggle (1975) or Cox (1975) must be used for this purpose.

The Height

The height is the vertical distance between the ground and the highest living part. For trees taller than 5 m an estimate to the nearest meter is sufficient; smaller trees should be recorded to the nearest 0.1 m. A measuring rod is the easiest method in low vegetation. In tall vegetation, where the top and the bottom of the tree can be simultaneously observed, clinometers can be used; otherwise any practical method can be applied. The mean maximum height for each vegetation layer (stratum) should be estimated from the data. For example, for a primary tropical lowland forest the emergent may reach 60 m, the canopy trees 40 m, the sub canopy trees 20 m and the under storey shrubs 6 m.

A Stem basal area

A stem basal area (A) is the cross-sectional area of the trunk (m^2) measured at the lowest point not influenced by basal swellings or buttresses. In forests it is traditionally measured at 1.3 m above ground level ('diameter at breast height', DBH), but in tropical forests the trees may still be buttressed at this height, and in savannas, branching frequently occurs below this height. The area can be calculated either from a diameter (d) measurement using calipers, or from a circumference (c) measure using a tape.

$$A=0.7854\ d^2 \text{ or}$$
$$A=0.0796\ c^2$$

Result will be in m- if the diameter and circumference are measured in m; if measured in cm, divide the answer by 10,000 to convert cm^2 to m^2. If the stems are very markedly non-cylindrical (for instance if they are eliptical or convoluted) an empirical correction factor should be calculated

A single plant has more than one stem, Where the area of each stem should be measured. For shrubs with very many small stems an estimate of the total combined area is adequate.

The total stern basal area for the plot is calculated by multiplying the stem basal area by stem density for each species (according to the criteria for species as defined above).

The Tree Biomass Estimation: Natural Vegetation

The cutting and weighing all portions of a tree estimates most precisely tree biomass. As this is very time consuming, tree biomass is more commonly estimated by statistical means, eg. by sub sampling, or on estimates based on tree attributes. When sub sampling is not possible, biomass is most commonly estimated from a regression equation.

As Since mass is the product of volume and density, tree biomass regressions often use diameter at breast height (DBH, cm), total height (H, m) and some measure of specific wood density (S, t/m^3). A locally developed equation is generally most reliable and should be used whenever possible. When no local equation is available, one of the regression equations given below can be used. These have been derived from several data sets and rigorously tested for reliability (Brown et al., 1989). Equations using D, H and S are the most reliable, followed by those using D and H, or just D. When H is not available one may apply double sampling methods, and estimate H as a function of D, then biomass as a function of D and estimate H. This is better than other approaches based on D alone when the H-D relationship accurately reflects the population of interest.

A regression relationships vary slightly with climate, so there are three equations reflecting three rainfall regimes; dry < 1500 mm, moist 1500-4000 mm and wet > 4000 mm. Not all equations are available for all climate types, due to lack of reliable data.

An equations may be applied to individual trees, mean trees or stand tables (numbers of stems per size classes) to estimate total above -ground biomass for a pixen tree, for a moan tree or per unit area respectively (Wharton and Cunia, 1985).

However, the following equations estimate biomass (B kg) and total height (H m) for trees in tropical forests. Values for the sample size, n, and goodness of fit, R^2 (an adjusted R^2) are included. The non-linear models include the adjustment term (MSE/2) together with the intercept, when appropriate.

An Estimation of Tree Biomass: Even-Aged Plantations

The Woody biomass estimation in even-aged plantations is simpler than in natural forest, and requires destructive sampling of many trees. The same principle of estimating biomass from predictor variable such as stem cross-sectional area and height, applies, but the model relating these variables to biomass cannot be assumed to be the same for different species and trees of different sizes.

In many species, for a given stand structure, the following linear function gives an accurate prediction of biomass (Stewart et al., 1992);

$$W = a_{ij} + b_{ij}\ \Sigma d^2$$

Where, W= biomass; D= stem diameter (Sd^2 being the sum of the squared diameters of all the stems of a multi stemmed tree); a and b are coefficients for species i on site j.

The coefficients have to be determined separately for each species and each site.

Forest type		n	R^2_{adj}
Dry	$B = 34.4703 - 8.0671\ D + 0.6589\ D^2$	32	0.67
	$H = \exp\{(0.671) + 0.6301\ \ln D)$	1231	0.54
Moist	$B = 38.4908 - 11.7883D + 1.1926D^2$	1680	0.78
	$B = \exp\{-3.1141 + 0.9719\ \ln (D^2H)\}$	168	0.97
	$B = \exp\{-2.4090 + 0.9522\ \ln (D^2HS)\}$	94	0.99
	$H = \exp\{1.0710 + 0.5677\ \ln D\}$	3824	0.61
Wet	$B = 13.2579 - 4.8945\ D + 0.6713\ D^2$	69	0.90
	$B = \exp(-3.3012 + 0.9439\ \ln (D^2H)\}$	69	0.90
	$H = \exp\{1.2017 + 0.5627\ \ln D\}$	69	0.74

It is established that although these equations have been derived from large (and sometimes very large) sample sizes, they should be viewed with some caution until validated for a given system.

Methodology

(i) Fell a minimum of 12 trees.

(ii) Partition the trees into components (wood of different sizes, levels, and pods/fruit).

(iii) Weigh the components immediately in the field.

(iv) Oven dry sub samples (>100 g) of each to determine water content.

(v) Calculate total dry weight (biomass);

The biomass and diameter(s) of the sample trees are used to determine a and b by regression analysis. If R^2 is low (<0.8) other models should also be tested. The biomass of the other trees in the plantation is then estimated from their diameters.

For a small stand this may be done for every tree individually; for large populations the total biomass may be estimated using the mean diameter and the total number of stems in the stand.

The twelve is the absolute minimum number of trees per species that must be destructively sampled, but if a large sample is used it will give a more reliable model with closer confidence limits. Whatever the sample size chosen, it is essential that the sampled trees cover the whole size range of the population. The usual procedure is to use a stratified sample in which the trees are divided into size classes on the basis of diameter (or cross sectional area in the case of multi-stemmed trees), and sample trees are selected randomly within each size class.

The studies on a single site, the inclusion of tree height h in the regression function is found to contribute little to the goodness of fit of the model. However, a linear model of the form

$$W = a_i + b_i h \Sigma d^2$$

is more likely to be applicable across sites than one using diameters only, because height, unlike diameter, can be assumed to be independent of stocking level and silviculture treatment (Philip, 1983).

The Herbaceous Plants And Short Duration Crops

The required measurements are total biomass (t/ha = 0.001* kg/ha=0.01 xg/m^2) and the contribution (%) of each species to that total. Total biomass is measured by harvesting, drying and weighing a number of small sub samples, or quadrats. Quadrat size depends on plants spacing; 0.5 m*0.5 m is a convenient size for most grasslands; 1 m x 1 m may be appropriate for crops such as maize. Sample number (n) should be sufficient to reduce the standard error to about 10% of the mean. 20-30 samples per treatment distributed amongst replicates is usually sufficient. Sample location should best be random, but systematic with a random start is acceptable.

Procedure

(i) Cut all herbaceous vegetation within the quadrat at 2 cm above the ground (to avoid soil contamination), and sort into live (green) biomass and standing dead if possible.

(ii) Collect the litter from the ground for an estimate of litter standing crop.

(iii) Dry all samples as soon as possible to prevent decomposition.

The Species composition in mixed communities is estimated by the dry-weight-ranking technique. The technique is based on a multiple regression for the dry weight of a mixed sample of herbage of the weights of the three heaviest species in the mixture. Experience indicates that it is easier to assess visually the rank order of the species in quadrat than to estimate accurately their biomass. Tests in a large number of different communities have shown that the regression coefficients are fairly consistent between communities, and therefore do not need to be recalculated each time. The technique does not work well in communities completely dominated by one species, and tends to ignore rate species. The modified form given here, where the total biomass within the quadrat is also given a visual score, gives better results in communities where the total biomass is patchily distributed. Quadrat size should be small enough that the species ranking is simple, but large enough that most quadrats have at least three species in them. Quadrats 0.5*0.5 m square are usually adequate in grasslands. About 50 quadrats should be assessed per treatment.

Procedure adopted

(i) Walk around the plot to obtain a clear visual impression of what the minimum (1) and maximum (5) quadrat biomass looks like.

(ii) Locate the quadrat randomly or systematically after a random start.

(iii) In each quadrat (i) give the total biomass a score (Wj) between 1 and 5 according to whether it is near the minimum or maximum for the plot.

(iv) In each quadrat, give the species (j), which contributes most to the total quadrat biomass a rank score (q;) of 1, the second heaviest species a rank of 2, and the third heaviest species a score of 3. If a single species contributes more than about 70% of the biomass, give it ranks 1 and 3, or 1 and 2 (or even 1, 2 and 3 if it is the only species in the quadrat). Similarly the second species could get a 2 and 3 if necessary.

(v) When all the quadrats have been scored and ranked, calculate the score for each species:

Calculation prescribed

Species j score $= 70.2f\Sigma\ W_i + 21.1\ r\Sigma W_i + 8.7r\ \Sigma W_i$

where, SWj is the sum of the quadrat scores for the quadrats where species j obtains rank r.

Add up the species scores to give a total score. Determine the % contribution by species j to the total biomass: Species j contribution to biomass (%) =(species score/total score) x100.

Example -1: Comparative Assessment of Methods for Estimating Biomass in Forest Ecosystem

Introduction

A marked increase in the number of studies on forest biomass during the past decades with the realization that total organic production is important instead of considering the forest production system for stem wood. This is probably due to increase in pressure placed on our forests by the community such as demands for forest products, search for renewable source of raw materials etc. In addition to the productive role, the growing concern of well being of forest ecosystems has resulted in the appearance of numerous publications on forest biomass throughout the world (Parade, 1980) including India (Singh and Sharma, 1976; Sharma 1971; George, 1977; Negi, 1984). These studies are utilized by some researchers with a geographical stand point, relating the distribution of productivity over land surface to either climatic factor (Paterson, 1956; Leith, 1972), or the distribution of main plant community (Rodin and Bezilevich, 1956). Such studies give us a general idea of productivity over the earth while neglecting the productive relationship in any particular forest stand. On the other hand some researchers are concentrating on the pattern and processes of production within a given forest stand as a base for understanding the energy flow and circulation of nutrients in the forest ecosystem (Satoo and Madgwick, 1982). Thus, biomass studies are essential for estimating Net Primary Production (NPP), understanding nutrient cycling and energy transfer and predicting the effects of tree utilization management procedures or other disturbances on productivity and stability of forest stand.

An oil crisis of 1970s has led a number of investigators to study the productivity of forests as a source of energy and chemical feed stock (Satoo and Madgwick, 1982). In India the man-made forests under Short Rotation Forestry (SRF), Social Forestry (SF), Energy Plantation (EP) and Farm Forestry (FF) program are of vital importance as these plantations indirectly help to relieve the tremendous pressure on the managed forest. However, it is essential to understand the potential nutrient drain from these forests under the wide range of harvesting practices including conventional clear-cutting.

The biomass estimation two distinctly separate situations can be considered. On one hand there is a typical research situation in which samples of trees and subordinate vegetation can be weighed in the field. While on the other hand the stand biomass can be estimated by prediction equations based on sampling over a wide geographical area. In this chapter, various methods of biomass estimation have been discussed and some comparisons have also been made between the results obtained by using different methods in a pure Sal forest (tropical moist deciduous forest) of East Dehra Dun Forest Division. The details of the stand are presented in Table 1.1.

Table A. Stand Details

1.	Area of the sample plot (ha)			0.10
2.	Age (years)			47
3.	No. of trees on the plot			80
	(i) Category	A	16	
	(ii) Category	B	21	
	(iii) Category	C	17	
	(iv) Category	D	14	
	(v) Category	E	12	
4.	No. of trees/ha		800	
	(i) Category	A	160	
	(ii) Category	B	210	
	(iii) Category	C	170	
	(iv) Category	D	140	
	(v) Category	E	120	
5.	Mean diameter of the crop (m)			20.0
	(i) Category	A	10.5	
	(ii) Category	B	15.0	
	(iii) Category	C	20.4	
	(iv) Category	D	24.4	
	(v) Category	E	28.9	
6.	Mean height of the crop (m)			19.6
	(i) Category	A	10.6	
	(ii) Category	B	15.2	
	(iii) Category	C	19.8	
	(iv) Category	D	22.2	
	(v) Category	E	23.8	
7.	Basal area (m^2)/ha			25.2

Proceduce adopted

The Selection of a sample plot: Generally it is not possible to measure all the trees of the forest stand, and hence the sample plots are taken to represent the forest. As such there is no firm basis for deciding on a suitable plot size but the smallest size of a rectangular plot should be large than the height of the dominant tree (Satoo and Madgwick, 1982). However, it will be possible to select a suitable plot size by trial and error method using some easily measurable stand parameter such as basal area (Freeze, 1961). The effect of plot size on variability of estimated weight is illustrated in Table 1.2 using data from a natural stand of Shorea robusta. All biomass components were more variable than conventional mensurational parameters of basal area, diameter (DBH) and height. It clearly indicates that at least larger or more plots should be used for the estimation of biomass,. Similar results were also observed in a stand of naturally regenerated *Pinus virginiana* (Satoo and Madgwick, 1982). Increasing the plot size will reduce the variation substantially (Table 1.2). Plantation would be expected to be less variable than naturally regenerated stands. Newbould (1967) suggested that sample plots in uneven aged mixed forests should be not less than 0.5 ha while in tropical rain forest the area must be 1/5 of the forest stand.

From the perusal of Table B indicates that the stand biomass is substantially affected by the plot size less than 200 m². Considering the total stand biomass the size of the plot should be around 1000 m², preferably a square plot of 30 m x 30 m since the variation in stand biomass becomes negligible. The reliability of the estimates depends on the sample trees felled; therefore, the judicious selection of sample trees is a first requirement. Forked and abnormal trees should be avoided and the trees on the edge of the forest should not be included.

Table B. Variability of stand parameters as affected by plot size in natural stand of *Shorea robusta*

Growth Parameter	**Plot size**			
	100 m²	**200 m²**	**400 m²**	**1000 m²**
(i)	**(ii)**	**(iii)**	**(iv)**	**(v)**
No. of trees	7	16	31	80
Basal	0.2015	0.476	1.163	2.5188
	(20.15)	(23.8)	(29.075)	(25.2)
Average DBH (cm)	19.0	19.5	21.9	20.0
Average height (m)	18.9	19.2	20.2	19.6

Biomass (kg)				
Foliage	42.31	122.54	242.99	600.95
Twig + Branch	164.37	530.40	1013.41	2457.23
Bark	219.59	550.54	1120.15	2959.79
Bole	819.19	2020.42	4290.35	10786.88
Total	1245.46	3223.90	6666.90	16804.85
Biomass (t/ha)	125.00	161.00	167.00	168.00

The tree parameters

It is always a high correlation between DBH, height and weight. Thus it becomes essential to measure the DBH of all trees in a sample plot and height of some of trees. For weight determination, the sample trees can be selected either from the plot or outside the sample plot. Regression of height and diameter can be used to estimate the height of the remaining trees; a free hand curve may also serve the purpose. For example from the height and diameter curve obtained from 21 sample trees, the following regression equation has been worked out.

$$Y = 4.8980 + 0.6832\,x \quad (r = 0.953) \qquad \ldots(1)$$

Y = Height (m) x = DBH (cm)

The Selection of sample trees

A Sample trees should be of good form avoiding forked or damaged tree. One should also keep in mind that the number of the sample trees for measurement is limited by manpower available. Tree weight is usually closely related to the size, increasing proportionally to diameter raised to the power between 2 and 3 (Satoo and Madgwick, 1982). Stratified sampling based on DBH can reduce the number of the sample trees; however to get a valid r jression model, it will be more useful if there is a fair distribution of sample trees covering the entire range of diameter on the plot. It is preferable avoid the trees growing on the edge in the natural forest.

Sample tree Measurement

Sample trees should be rneasured before felling in an identical way to those trees within sample plot for which weights are to be estimated. These trees are selected within or outside the sample plot depending upon the availability. In case of present study 19 trees of different diameters were harvested from the stand located in a pure Sal forest (tropical moist deciduous forest) of East Dehra Dun forest division. To develop good regression model 18 more trees were harvested from different localities. Thus a total of 37 trees distributed over the entire diameter range from 9.1 cm to 37.1 cm were harvested.

All the tree components viz., foliage, twig (2 cm), branch (2 cm) and bole were separated immediately and their fresh weight recorded in the field. A representative sample of each tree component (100 g for foliage twig and branch) was also taken for its oven dry weight estimation (80 °C). The bole portion of sample trees was cut into 2 m or I m long sections for the convenience of weighing. A 5 cm wide disc was also removed from the base of each billet for the estimation of fresh and oven dry weights of bark and wood (under bark) and also for the estimation of volume (over bark and under bark) of the main bole (up to a diameter limit of 5 cm over bark).

There is a marked variation in the heights of sample trees. As the diameters of the trees increase, there is a corresponding increase in the heights also for a particular site. Thus the site effect becomes conspicuous and is reflected in better height growth for the good sites, which in turn depends on the stand density also.

Estimation of stand weight for sample trees

The view given by Satoo and Madgwick (1982) the following steps are involved in estimating the stand weight:

(i) Selection of sample plot.

(ii) Measurement of limited number of dimension on all standing trees in the plot.

(iii) In the case of research plots of selection, detailed measurements of sample trees.

(iv) Application of values for sample trees to the remaining unit area.

There are several general categories of methods used to estimate the stand weight from sample tree data.

Mean tree technique

The method is based on the detailed measurement of one or more trees of mean dimensions. Sample tree data are converted to a land area basis by multiplying the biomass of different components by the total number of trees per unit area. The average biomass in different components of three mean trees is used to get the biomass on unit area basis.

Stratified tree technique

This method is almost similar to the mean tree technique. Different diameter classes stratify the stand. One or more trees are harvested from each diameter class having parameters close to the mean of the class and the biomass figures of different components of each stratum are added up to get the biomass on unit area basis.

Unit area method

This method involves the harvesting of all the trees standing on a randomly selected plot. Depending upon the size, age, diversity, availability of manpower etc., this method is too much time consuming and calls for much hard labour and therefore, seldom used in biomass studies.

Basal area proportion method

In this method a tree of mean basal area is harvested and the weight multiplied by the number of trees per unit area; this gives the biomass. When a tree of mean basal area is not available, several trees closest to the mean basal area are sampled and the following formula is used for biomass estimation

$$W = \Sigma_W \frac{G}{g}$$

where W is the weight of trees of the plot, vv is the sum of the weight of sample trees, G the sum of basal area of the plot and g the sum of basal area of the sample trees. A better estimate may be obtained when stratification is made with the diameter and taking trees of mean basal area of each stratum (Satoo and Madgwick 1982).

Regression estimation technique

This method is also called "dimension analysis of woody plot" (Whittaker. 1961) and "allometry" (Kira and Shidei, 1967). This technique involves three steps as discussed below:

(i) Trees are tallied by species and diameter at breast height,

(ii) The weights of the sample trees are related to easily measured parameters (DBH, heights etc.),

(iii) Application of regression obtained in step (2) to step (1).

The regression equations for biomass prediction of different tree components have been obtained from 35 sample trees, which are given in Table (Negi, 1984).

Table C. Correlation coefficients (r) and prediction equations for different tree components as well as total biomass in *Shorea robusta* (Negi, 1984)

Dependent Variables. Y	Log DBH (x)		LogD²H (x)	
	Regression equations	r	Regression equations	r
Foliage	log y= -2.28061+ 2.33526 log x	0.944	log y= -2.52895 + 0.84866 log x	0.935
Twig	log y= -2.26097+ 2.23036 log x	0.930	log y= -2.48657+ 0.80640 log x	0.920
Branch	log y= -2.42358+ +2.86678 log x	0.936	log y= -2.67889+ 1.02741 log x	0.16
Bark	log y=- 1.403 74+ 2.25082 log x	0.954	log y=- 1.78422+ 0.85390 log x	0.991
Bole	log y=- 1.95385+ 3. 10398 log x	0.986	log y= -2.34087+ 1. 14144 logx	0.993
Total	N.D.		log y=- 1.78825 + 1.04848 logx	0.994

Root	log y=- 1.720 10+ 2.65696 log x	0.977	log y= -1.88750 +0.93460 log x	0.981
Total	N.D.		log y=- 1.54325+ 0.93460 log x	0.995

N.D. = Not determined

An Out of the five methods used, the mean tree technique and the basal area proportion method gave six to eight per cent under estimates of the total biomass of 166 t/ha estimated by stratified tree technique. Results from the test plots reported by Satoo and Madgwick (1982) indicate that the mean tree method can give under estimates of canopy weights; however, estimates based on several trees closest to mean basal area generally gave results within ten per cent of actual plot values. The frequency distribution of diameter classes may not necessarily be symmetrical about the mean and the asymmetrical distribution may result in the under estimates or over estimates of the biomass. In the present case, the best results have been obtained by regression methods after the stratified tree technique. Since there is a good correlation between the tree weight and the linear size (DBH, height etc.) the regression techniques are most widely used and the only methods for estimating biomass of the stand where no direct sampling has been undertaken though the component weights particularly the leaf is subject to large variation (Satoo and Madgwick, 1982). In our discussion the regression estimates of foliage weight as compared to that of stratified technique are about 16 per cent under—estimations while the (twig + branch) estimates are within 6 to 8 per cent. The stratified tree technique good and reliable estimates if sufficient numbers of trees are included in each stratum; the regression technique is the next choice where destructive samplings are not possible.

The stratified tree technique is undoubtedly the best method but it is time consuming and expensive; therefore, there is a need to develop estimating equations of wide applicability so that reliable estimates of stand biomass may be obtained without recourse to the laborious destructive sampling.

A comparison of biomass estimated by using various methods has been made in Table D

Table D. Comparison of biomass (t/ha) estimates obtained by using various methods in *Shorea robusta stand*

Components	Mean tree Technique	Stratified technique	Basal area proportion Method	Regression method LogD²H	LogD
Foliage	5.3	5.7	4.5	4.8	4.7
Twig+Branch	15.5	23.5	20.1	22.1	21.6
Bark	27.6	27.2	24.7	47.7	27.4
Bole	107.6	109.6	104.0	112.0	112.3
Total	156.0	166.0	153.3	166.6	166.0

Example - 2: Overview of Methodology for Teak Biomass and Nutrients Study (Within Each Climatic Zone)

(i) Select 5 plantations to cover wide range of ages, trying to keep climatic and soil factors constant. This is a chronosequence.

(ii) For each plantation, randomly locate and layout about 20 plots, attempting to ensure that they cover the range of site within the plantation. Plots should contain a minimum of 12 trees - it is important to be as small as 10 x 10 m.

(iii) Measure the girth at breast height of all trees, and record these and other site details on the standard form.

(iv) Calculate the basal area (BA) and stocking per ha

(v) Group the 20 plots into 5 groups on the basis of BA (perhaps taking into consideration stocking or mean girth).

(vi) Select the mean plot from each of five groups - these will be sample plots to be used for the measurement of the biomass trees.

(vii) Take soil samples from each of the 20 plots, using the depths 0-20, 20-50 and 50-90-cm. Label samples with plot number and soil depth. Analyze using the standard methods.

(viii) Rooting depth has to be estimated for each of the 20 plots.

(ix) Soil description should come from one or more pedons which represent the soils in the compartment.

(x) KT. Bulk density for the three depths used for the above sample analysis needs to be measured also. Bulk density for the 0-20 cm is likely to be more variable but more critical, so for this horizon it should be measured in each of the sample plots as well as at the pedons.

(xi) Biomass of one tree of about average girth from each of the 5 selected sample plots. This tree should be typical of the plot and not grossly malformed.

(xii) Follow the standard destructive biomass procedure and enter the data on the field form provided.

(xiii) Enter the data on the spreadsheet developed for this study, backing up the data on computer floppies (at least two backups).

(xiv) Use the modified basal area ratio method to calculate the per hectare values for each of the 20 plots.

(xv) The data set for subsequent analysis will be:

For the matrix of 5 plantation ages and 20 plots per compartment, there will be biomass by tree components, nutrients by tree components; soil nutrients by various depths, rooting depth, tree volumes, basal area and stocking.

(xvi) Data analysis could employ the following techniques:

Analyses of covariance to see if there are significant stand age differences.
Covariate could include a composite soil/site factor.
Correlation techniques.

Regression models.

Other multivariate techniques.

Graphical methods.

Balance sheet approach.

Effect of harvesting practices on nutrients.

Site description sheet for biomass studies in teak plantations

Plantation number : Planting Year :

Sampling Date : Age (Years) :

Division : Range:Compartment :

Stocking (stems/ha) :

Soil type :

Topography :

A Sample plot data sheet for biomass studies in teak plantation

Plantation Number : Plot Number :

Plot size: m = m^2

Aspect Slope(degrees): Topo. position

*Top, Mid, Lower or bottom

1		12		23		34	
2		13		24		35	
3		14		25		36	
4		15		26		37	
5		16		27		38	
6		17		28		39	
7		18		29		40	
8		19		30		41	
9		20		31		42	
10		21		32		43	
11		22		33		44	

Record all girths to nearest mm. eg 24.4 cm; correct plot areas for slope.

Summary :
Stems/ha :
BA/ha. (m^2) :
Mean girth (cm) :
Soil Sample collected? : Y/N
Signature:

Teak biomass sampling for procedure

The introduction of the new standard forms and then entry onto the spreadsheet should lead to fewer errors and higher quality data. The system has been designed for ease of use.

Procedure observed in Field

(i) Before felling the tree, measure crown width (2 diameters), girth at stump, BH and on large trees or fluted trees, the girth at about 0.6 -0.7 m. Record these to nearest mm, and mark breast height point.

(ii) Fell tree, preferably onto a sheet.

(iii) Lay tape along the tree using the BH mark (1.37m) as a reference point. Enter the height at which the stump was measured and the stump disk is to be taken.

(iv) Measure height to first branch and total tree height (all heights to be recorded to nearest 0.1 m).

(v) Decide on how many growth measurements and stem sample discs need to be taken for good representation of the stem. The form allows up to four discs to be taken. Discs should be taken at approximately equal distances along the trunk. There will always be at least two discs (stump and top disc). Mark these points on the stem, recording their heights.

(vi) Cut the stem and weigh the logs, recording to the nearest kg.

(vii) Immediately cut the sample discs (uniformly 5 cm thick) before drying can occur and quickly weigh these (with bark) to nearest g. Label the discs with the sample number; record measurements and label number.

(viii) Where a disc is taken, girth, bark-thickness and sapwood thickness should be recorded (nearest mm), taking the mean of 4 measurements.

(ix) Complete the menstruation data on the stem by measuring additional girths and bark thickness, as may be required.

(x) Note: volumes OB and LIB of the stem will be calculated by Smailian's formula, which require the girth at each end t-f the log and the log length (and the bark thickness to get UB volumes).

(xi) Branch wood (>20cm girth) should be weighed green, length measured to nearest 0.1m and the single mid-point girth taken. Record each branch wood separately on the form. (Calculations use Huber's formula for branch wood OB volumes).

(xii) Quickly sample by taking a series of 2-cm thick discs (say 2 or 3 from each branch wood) from these branches, covering the range of diameters. Combine together into a single sample; quickly weigh to nearest g, label and record weight and label number.

(xiii) Do not attempt to take a uniform 100 g sample.

(xiv) Follow a similar procedure for branches, twigs and leaves, combining many small randomly chosen pieces to make a sample.

(xv) For branches, twigs and leaves, total fresh weight is recorded.

(xvi) Remove the roots and divide into the stump and central core region (say down to 20 cm girth) and into coarse lateral or sinker roots (> 2 mm). Weigh and record these two separately.

(xvii) The stump portion is not sampled but the stump disc taken at the bottom of the log is assumed to represent it. Coarse roots are sampled, by taking several pieces at random to represent all sizes.

(xviii) Before leaving the tree check that all data have been recorded as required, and that all samples are clearly labeled and are being taken to the laboratory.

A procedure adopted in laboratory

In the laboratory there will be separate forms that will have dry weights and nutrient analysis recorded on them. The following points should be noted:

(i) Labeling should be clear and will have been recorded on the field form if there is any query. Check that all samples are present.

(ii) Stem discs will have to be divided into wood and bark (greenweights of each not required) and need to have further labeling (e.g. a W or B added to the number

(iii) After drying all samples at 65-70°C until there is no further reduction in weight, weigh and record to the nearest g (nearest 5 *or* 10 g is not good enough). There are sheets for recording these.

(iv) Finely grind and analyse for N, P, K, Ca, Mg, and Na concentrations; record on appropriate sheet.

(v) Enter all data from the field and laboratory sheets into EXCEL using TEAKCAL.XLS spreadsheet. This will do all the calculations. Check that data.

(vi) Place a hard copy of the summarized data on file.

Measurements conducted

	Units	Measure and record to nearest
Crown diameter	m	0.1m
Heights	m	0.1m
Girths	cm	mm

Bark thickness	mm	mm
Sapwood depth	mm	mm
Fresh weights	kg	kg (or within 3 %)
Sample weights	g	g
Nutrients	%	varies

Sample tree field data sheet for biomass study of teak

Plantation number: Date sampled:

Plot number: Tree number:

Crown diameters (m) 1. 2.

STEM

	Height m nearest 0 .1	Girth cm	Bark thickness mm	Sapwood thickness mm	Total FW kg	Sample FW G	Sample No.
Stump	0.						
	0.						
B.H.	1.37						
Top							

Notes: There must be at least two discs (stump and top) taken; others may be taken if dealing with large tree. Enter the total fresh weight of each section before cutting the disc. For accurate volumes additional girth and bark thickness (mean of 4) measurements are required: at about 0.7m and at BH, and at about 2-3 m intervals above breast height on trees with long boles. The average sapwood thickness (mean of 4 and excluding bark) is required for each disc.

Height first branch (m):

Total tree height (m):

FOLIAGE

Total Foliage F W (kg):

Sample F W (g):

Sample no:

BRANCHES

Branch wood				**Branches**		**Twigs**	
Length M	Girth cm	FW kg	FW Sample G	FW kg	FW Sample G	FW kg	FW Sample G
			Sample No.		Sample No.		Sample No.

Notes: - Branches are divided into three categories :> 20 cm girth; 5-20 cm and twig (<5cm). Large branches are freshly weighed separately (to nearest 0.1 kg); other branches or twigs may be weighed together. Sub samples are required from each category and should be weighed green in the field to the nearest g. Branch bark need not be separated from branch wood. For the branches >20 cm their girth at the mid point along the branch and length is required to obtain branch volume (Huber's formula).

ROOTS

Stump F W (kg):

Coarse root (kg):

Sample Coarse root FW (g): Sample no :

CHECK LIST

All field measurements entered Y/N

Number of samples? .....

Samples clearly labeled? Y/N

Measured and recorded by: .....

Biomass sample sheet for recording dry weights

Plantation no:

Plot No:

Tree No:

Stem samples

Field Sample no.	Fresh Wtg.	Wood dry wtg.	Bark dry wtg.

Note: The wood and bark sample numbers will need to be given either a/W or a/B to distinguish them apart once they have been separated and prior to analysis.

Other Sample

Sample no.	Type	Fresh wt. g	Dry wt. g
	Foliage		
	Branch wood		
	Branch		
	Twigs		
	Coarse roots		

Dry the samples until they reach a constant weight and record the values in the table. The fresh weights are recorded here again for checking purpose

Time : Result recorded by:

Date :

2

A Technique for Nutrient Cycling

General Introduction

An interchange of chemical elements between living organisms and the physical and biological components of forest ecosystem may form an extremely complicated system, essentially cyclic in nature and may broadly follow the organic system (Ovington, 1962). In order to understand the magnitude and pattern of circulation of the chemical elements, which may be characteristic of an individual element, the cycling rate and standing state will have to be quantified. Nutrient cycling may be a good indicator of the continuity and stability of any living system and may be believed to be a useful strategy for ecosystem analysis (George, 1977)

A system has been constitute by forest and soil together which work complementary to each other. The basic mineral nutrients to the forest crop are provided by the soil due to weathering of rocks and minerals. The soil is recharged with the mineral nutrients after the decomposition of the organic matter obtained from trees due to leaf fall, etc., the nutrient cycle in forest means movement of basic nutrients from atmosphere to soil and soil to the atmosphere. Several factors such as water, micro organisms, condition of decomposition of organic matter etc., affect these cycles.

In a forest ecosystem, two major pathways of flux can be recognized namely:

(i) Bio-geo chemical (open) cycle.
(ii) Biological (closed) cycle

The bio-geo-chemical consists of a nutrient reservoir pool, which is large and there is exchange pool which is small but active. The biochemical cycles involve gaseous nutrient cycle, where atmosphere is reservoir pool and sedimentary nutrient cycle, where earth crust is reservoir pool. These cycles involve input and output of mineral elements. The inputs of nutrients include:

(i) Addition of nutrients by precipitation
(ii) Weathering of parent rocks
(iii) Nitrogen input through nitrogen fixation by micro organisms, and electric discharges in thunder storms.

Precipitation is one of the most natural means by which nutrients are added to the forest ecosystem. It is reported that the annual nutrient supply through rain water in temperate regions are less compared with tropical regions.

Supply of nutrients is also made in soil by weathering of soil particles and parent material. This is a part of sedimentary cycle. Weathering of soil particles increases nutrient availability and weathering of rocks increases nutrient capital in the soil. The rate of weathering is high in forest soils, because of better weathering environment. Nitrogen is added in soils by symbiotic and free living micro organisms. The net annual gain of nitrogen may be about 50kg to 500kg per ha in tropical forests. Forest canopies also trap air-borne materials.

The main sources of output are loss of nutrients through harvesting of trees, grazing, leaching, fires etc. The biological cycle may involve a cyclic circulation of nutrients between the forest soil and the biotic communities and may include the processes of mineral uptake (or) absorption, retention (in the annual accumulation of biomass) and restitution (or) losses (leaf litter, organic debris, washing by rain etc) (Duvigneaud et al., 1970). This may be expressed as below:

Nutrient uptake = Nutrient retention + Nutrient returned

In this chapter, the methodologies for investigating the nutrient cycling in forest and plantation ecosystem are described. To study the nutrient cycling, nutrient retained and nutrient returned should be calculated.

Retained Nutrient

The nutrient retained in the annual biomass production of forest vegetation may be obtained by multiplying the biomass components with their respective nutrient concentration. The nutrient contents of the standing crop may differ substantially between various forest types according to the degree of uptake and crop biomass, which may in turn depend on tree age and site quality, even within the same species. The increase in nutrient contents of the standing crop with stand age observed in plantation and forest stand should have a direct relationship to the total biomass (Rodin and Bazilevich 1967). As such, the nutrients accumulated in various tree components may vary considerably.

Return Nutrient

Estimation of litter production and its nutrient content per unit area needs to be studied to investigate the nutrient return in an ecosystem. Besides nutrient in litter fall, the forest floor also receives them from through fall, stem flow and canopy interception.

Procedure to Study Nutrient Cycling

Estimation of Litter

Litter is conveniently divided into two components, fine and coarse litter. Fine litter is mainly leaves but also includes flowers, fruits and fine twigs. It is relatively easy to set a system of traps, for example fine nylon-mesh trays, and to obtain a statistically satisfactory record of fine litter fall over a year (or two) of study.

Coarse litter, which is branches, limbs and falling tree trunks, is very heterogeneous in both space and time and it is much more difficult to get a meaningful record, although in amount coarse litter may exceed fine litter.

Construction of Litter trap

Litter traps are bags or boxes supported just clear of the ground with an aperture of 0.25-1 nr. A circular construction is best as it minimizes edge effects. Woven plastic bags are light weight for use in remote sites and can be tensioned into shape using lines attached to D-rings sewen around the mouth of the bag. The traps must allow free drainage of rainwater but have a mesh size of approximately 1 mm or less to retain fine litter fractions. Trays on the ground surface can be used to measure litter-fall from dwarf shrubs etc.; but animal activity, drainage and wind can present problems.

Similar considerations apply to collections of litter from quadrats on the ground, which may be necessary for estimating falls of palm fronds and large woody litter. Trash fractions, which often have low mass but high nutrient content, however, will be lost by this method.

Methodology adopted

(i) Randomly locate litter traps (for material other than branches) within moderately homogeneous plots, or in a stratified random pattern (with 10 traps per subplot) in sites where it is necessary to include major variation in topography, soils and vegetation structure. Note: To achieve a 5% standard error about the mean, Newbould (1967) recommends the use of a minimum 20 traps/plot. In very heterogeneous sites, however, higher numbers of traps may be required.

(ii) Collect litter every 2 weeks and air-dry it. More frequent collections may be necessary for litters, which decompose rapidly, e.g. some tree legumes and while less frequent collections may be made under dry conditions (though the possibility of the litter becomes contaminated with dust and or animal faeces should be recognized).

(iii) Son the dried material into:

a. Leaves (including petioles and foliar rachis);
b. Small woody litter (twigs < 2 cm in diameter and bark);
c. Reproductive structures (flowers and fruits could be differentiated);
d. Trash (sieve fraction < 5 mm).

(For palm fronds, the leaflets, the rachis below 2 cm, and the remaining parts of the rachis should be weighed and recorded separately).

(iv) Oven-dry sub samples of litter to obtain correction factors for moisture content

(v) Express all fractions defined above on dry basis in g/m2/year or t/ha/year with 95% confidence limits.

(vi) Estimate branch fall from large (e.g. 100 m2) ground quadrats. Break twigs at the 2 cm diameter point, weigh the > 2 cm diameter material, sub sample for oven-dry mass and other determinations as required.

Above-ground crop residues and Herbaceous litter

The minimum level of sampling is at maximum and minimum biomass associated with major seasonal changes "or perturbations; i.e. sampling four times a year under climatic regimes with strongly bimodal pattern of rainfall. This will underestimate litter inputs as a consequence material turning over between sampling dates and sampling at regular intervals every few weeks is recommended.

Methodology & Procedure

Determine herbaceous litter (including grasses and forest ground flora) by harvesting quadrats in conjunction with biomass estimates.

Separate litter, where possible, by plant species for the most frequent 80% of species and bulked for the remaining 20% (This may be impractical in species-rich communities).

Determine crop residues after harvest and at the time of ploughing or other manipulation.

Oven-dry litter sub samples to obtain correction factors for moisture content. Litter heavily contaminated by soil may need to be corrected for 'ash' content as well.

Express all fractions defined above on an oven dry basis in g/m2/year or t/ha/year with 95% confidence limits

Requirements of Other inputs

A Mulches, manures and fertilizer

Information on the type of input applied, the amount applied, and the date of application is necessary. For example, information required for mulches would include the type of mulch (e.g. rice straw), the amount applied on a dry weight basis (t/ha), and the application date(s). A representative sub sample can be taken for nutrient and other analysis. Details on methods incorporation (e.g. surface application, ploughed to a certain depth, banded) of each of the inputs should also be noted. Similar information is necessary for manure application.

For fertilizers, specify the material applied (e.g. urea, single super phosphate), quantity applied and application date (s).

Herbivory

Record the dates of commencement and termination of feeding, the type of herbivore (species, browser, grazer, etc.), and the number of animals in the plot and estimates of their individual masses. These data will permit the stocking level (kg/ha) to be calculated, given the plot area, and the rates of plant consumption, defecation and urination (kg/ ha/day) to be estimated.

FLITTER DECOMPOSITION AND NUTRIENT RELEASE

Decomposition is a complex process regulated by the interactions between organism (fauna and micro organisms), physical environmental factors (particularly temperature and moisture) and resource quality (lignin, nitrogen and condensed and soluble polyphenol concentrations) (Swift et al., 1979). As decomposition progresses, soluble and particulate materials from the litter, organism tissues and products of microbial metabolism are separated from the original resource by leaching, physical fragmentation and animal feeding activities. These products are then transported by wind, water and gravity to soil microhabitats, which have a different set of conditions regulating decomposition to those of the parent material. An almost intractable problem in quantifying litter decomposition is the need to impose methods, which enable the experimental material to be identified without affecting the variable, which regulate the component processes.

Litter bags and decomposition rate constants

Enclosing litter in a mesh bag makes it possible to recover the residual experimental material and defines the conditions under which the organisms operate. However, the mesh bag and compaction of the litter can create different microclimates to condition in unconfined litter and the incorporation of material into soil is also affected. Nonetheless the standard litter bag remains one of the most convenient methods for comparative measurements of decomposition between sites and treatments.

The minimum requirement for litter decomposition rates in these studies is the time for 50% loss of the initial mass (t<-0) but most studies in the literature report the decomposition rate constant "k" from the negative exponential function:

$$W_t / W_O = W_O\ e^{-}kt$$

Where W_t is the amount of the initial mass (W_O) remaining at time "t".

This is more conveniently expressed as the regression function:

$$\text{Logn}\ (Wt / W0) = \text{logn}\ W0 - kt$$

The use of the single exponential model assumes that decomposition rates are constant over time. In fact, litter mass losses often show a fast initial phase due to leaching of water-soluble material (which is largely an experimental artifact of drying and re-wetting the litter), a slower second phase dominated by the decomposition of cell wall constituents such as hemicelluloses and cellulose, and a slow phase regulated by lignin and microbial products. This series of curves can be described by fitting a two - (or) three - phase exponential model with a rate constant for each of the fractions (Wieder and Lang, 1982).

A Decomposition of unconfined litter and turnover rates

A Measurements of mass losses from unconfined litter under natural conditions are only possible in the initial phases of decomposition before the material starts to disintegrate. The method is most satisfactory for wood and other discrete materials, which can be tagged or attached to a tether (Lang and Knight, 1979). For branches and tree boles changes in relative density (weight per unit volume) provide a useful measure of mass loss without sacrificing the whole unit. Extensive decomposition of wood often occurs before branches and tree boles fall to the ground. This can be determined by comparing the relative densities of standing dead and forest floor material with live wood but the age of the material is difficult to calibrate (Swift et al., 1979).

Litter decomposition rates on the soil surface can be estimated from changes in litter standing crops over intervals of time where (i) there is a pulsed input of material (such as crop residues at harvest), (ii) the material has decomposed before the next input, or (iii) age classes of material can be distinguished. In most natural systems it is often difficult to distinguish litter cohorts; (e.g. grasses). Under these conditions determinations of litter turnover rates provides a useful basis for comparing litter dynamics in different habitats in relation to climate and site factors. The turnover coefficient "k" is calculated as the quotient of litter input (1) divided by the standing crop (S):

Even for small litter, the same fractions must be represented in the inputs and standing crops when calculating the quotient and this is usually only possible for leaf litter and small twigs (Anderson and Swift, 1982). Sites subject to fires (Woods and Raison, 1982) and heavier grazing present obvious difficulties in estimating litter turnover by input/standing crop determinations.

It is unfortunate that, for historical reason, the turnover-coefficient and decomposition-rate constant, which have no formal relationship, are both designated by the letter "k" and the use of this term must therefore be specified. **K = I/S**

A Litter bag methods observed

Modified method taken

For studies in agricultural systems, where inputs of crop residues, green manures or dung have largely decomposed by the start of the following cropping season, an alternative procedure has been developed using bags as mesh cylinders to retain the experimental material in contact with the ground. The method is most appropriate for surface litter treatments; sampling and processing buried litter requires modification of the protocol depending on the material and soil texture.

The advantages of the method are that it reduces microclimatic artifacts, allows natural access by soil fauna, enables measurements to be made of litter incorporation into soil and sampling can also be combined with determinations of other soil properties and processes. The disadvantages are that the method involves disturbing the soil (hence is most appropriate for arable systems) and samples take longer to process. The method is also less suitable where litter fall occurs during the experimental period though this can be excluded by fine mesh covers (e.g. mosquito-net). The method is unsuitable where there is residual standing crop of litter.

Procedure

(i) Open the bags to form mesh tubes (ca. 20 cm diameter by 30 cm deep) and bury them to about half of the depth in the soil. For stony or shallow soils the bag can be cut in half and only 5 cm inserted below ground. Placement is made easier by using a metal sleeve, which fits, inside the bag to hold the cylinder rigid during handling. In tilled soil, large stones may be removed from within the cylinder to facilitate subsequent sampling.

(ii) Locate the mesh cylinder on a random basis in the field and add a known mass of experimental material in parallel with plot treatments.

(iii) Sample 3 to 5 bags on a random basis at regular intervals. Collect as much of the litter from the soil surface as possible (this may include small stones and aggregates) and put into a bag.

(iv) *Surface-litter treatments:* Soil cores (ca. 5 cm diameter) can be taken after the removal of litter and divided into depth increments to determine moisture content, organic carbon distribution, soil organic matter fractions and N-mineralization rates in the laboratory. The remaining material can be dug out and hand sorted for roots and soil macro-fauna.

Buried-litter treatments: Take cores for associated measurements and then remove all the soil in the bag to a suitable location for processing.

(v) Fill in the hole with soil from the margin of the plot and mark the location to avoid resampling.

(vi) *Surface-litter treatments:* In the laboratory sort the bagged litter material by hand, or shift in a coarse sieve, to separate the large litter fragments, stones and soil aggregates such as termite/earth worm worked material.

Place the residual litter and soil material in a bucket of water, briefly swirl to suspend litter fragments, and carefully decant through a 2mm mesh sieve. Experimental material retained on the sieve is classified as litter. Use a 0.25 mm sieve to further separate material in the washings if soil litter is to be determined.

Buried litter treatments: Hand-sort the material for litter (fauna and roots) as far as practical. Sub sample the soil and separate by water floatation as described above. In soils with high clay content, sub samples may be deflocculated with a water softener to release occluded material

(vii) Drain washed litter thoroughly before air-drying. Take sub samples for oven dry weight determinations.

Note: Little leaching occurs if washing is brief but this can be readily checked. It may be impractical to use distilled water for this separation but the water chemistry should be borne in mind in subsequent analysis. Litter which is heavily contaminated by clay material may be separated by more prolonged washing but effects on nutrient leaching must be assessed. Water softeners used in deflocculating may have high concentrations of phosphorus, so samples treated that way may not be suitable for nutrient analysis. In some cases, such as where termites have in filled woody materials with soil, it may be necessary to ash the experimental material and express results on an ash-free basis.

Analysis of data

Plot the data for % initial mass, and logn % initial mass, vs. time (days) to inspect for trends.

If the untransformed data show linear trends, estimates of 150 are made by calculating the regression and solving the equation for t = 50%.

The convention is to fit a single negative exponential function to derive the decomposition rate-constant "k" ; 150 can then be estimated from the reciprocal of k/day.

The rate-constant "k" is the slope of the regression of logn % remaining vs. time. If time is in days then the slope is -logn % day. To obtain k/year multiply by 365.

For example, for *Shorea affinis* (sampled over 294 days in a rainforest in Sri Lanka), k = -0.0036/day; k/year = 0.0036 x 365 = 1.22/ year (values for k are conventionally expressed as positive values). The estimate of t50 = 1/0.0036 = 278 days.

Decomposition standards

Birch *(Betula spp.)* lollipop sticks are used as decomposition standard for inter-site and regional comparisons using natural vegetation (where possible) and a derived site in each study area.

The sticks (100 x 10 x 1 mm) are secured in bundles of three with a wire or plastic tie at each end of the bundle. Two sets of bundles are used to investigate the effects of litter and soil conditions on decompositions rates.

Procedure

(i) Place one set horizontally on the soil surface and bury the other set upright in the soil with the tips at the soil surface. Pairs of bundles, one of each set, are placed and samples on a random basis.

(ii) Collect sample of 10 bundles of each set at intervals of 3 months.

(iii) Remove the middle stick from the bundle, dry at 85° C and

A Soil and Plant Sampling for Laboratory Analysis

Sampling and sample preparation is arguably the most important stage of any study; excellent laboratory analysis does not compensate for poorly collected, poorly prepared or unrepresentative sample. Wherever possible, label all sample containers prior to going to the field. Field-fresh soil samples are often moist; these should be collected in plastic bags or tubes rather than paper bags. Plant materials, on the other .hand, should be collected in paper bags to reduce decomposition accelerated by a moist environment prior to preparation for analysis. Upon return to the laboratory, examine plant samples for contamination, e.g. dust or mud; a brief rinse with water may be required.

Unless the analysis requires a field moist sample, air-dry all samples as soon as possible to halt biological transformations.

A Soil sampling technique Auguring procedure observed

(i) Remove surface plant and litter material from a minimum of 5 randomly selected auger sites.

(ii) At each site auger to a known depth (e.g. 30 cm) and put the auger contents in a clean plastic bucket so that all 5 auguring can be thoroughly mixed prior to sub sampling. Take a sub sample at about 500 g (one cupful). Discard the remaining soil.

(iii) Repeat step 2 for the next depth required for each auger hole.

Undisturbed cores procedure adopted

(i) Clean a flat surface (either a vertical face in a profile pit or a horizontal surface at any known depth).

(ii) Firmly drive a soil coring cylinder (of known volume and mass) fully into the face causing minimum disturbance and compaction to the core.

(iii) Cut away surrounding soil as to be able to cut neatly across the inserted end of the cylinder.

(iv) Carefully lift the core away, trim the ends flush with the cylinder ends, and cap tightly.

Plant sampling taken

The part of the plant that needs to be sampled and the sampling strategy depends on (i) the purpose for which the samples will be used (i.e. the types of analysis), and (ii) the type of plant to be sampled. No standard method is thus applicable. The general concept is taking several (minimum 5) sub samples, mixing them to make a composite sample and finally taking a sub sample for analysis.

Soil, Plant and Litter Material Preparation

Soil material taken

Note: As many of the variables of interest in ecological studies are biotic, or biologically medicated, several of the analysis must be conducted on field-fresh soil samples; the following guidelines relate to soils that do not need to be field-fresh.

Procedure obsered

(i) Air-dry the soil by spreading it out in a shallow tray in a well-ventilated place protected from rain and contamination. Alternatively soils can be dried in a forced air oven at a maximum temperature of 60°C. Break up any clay clods. When the soil is dusty it is dry enough.

(ii) Crush the soil lumps gently so that the gravel and roots etc. are separate from the mineral soil.

(iii) Sieve the soil through a 2 mm sieve leaving the gravel and roots etc. in the sieve.

(iv) Pick out the roots and save if required.

(v) Retain the gravel for weighting if required. This should be done if it appears to be about 5% or more by mass of the original sample.

(vi) Retain a representative sample (of the sieved soil approximately 250 g), by e.g. coning and quartering, for analysis.

Note: Coning and quartering is a simple and effective method of sub sampling loose, <2 mm material: mix, then pile the material in flat cone or mound. Divide into four quarters by cutting across the middle in an "X" shape with a blade. Discard two opposite segments, and mix and re-cone the other two. Repeat the process until sufficient material remains.

Plant material and required

Procedure adopted

a. Dry green material and litter at a maximum of 60 C; dung should be dried at a maximum of 40 C to prevent volatilization.

b. Grind all the material to pass a 0.15 mm mesh.

c. Retain a representative sample of approximately l0g, by e.g. coning and quartering, for analysis.

Sample storage technique

Both soil and plant material will satisfactorily store for long periods (i.e. > 1 year) if they have been correctly prepared. Samples are stored in clearly labeled, air-tight, termite-proof containers; the storage room should be well ventilated.

PRESCRIBED EXAMPLE: NUTRIENT CYCLING STUDY IN A TEAK PLANTATION: CONCEPT ADOPTED

The purpose of this example is to look at the major nutrient pools and cycles in a semi- mature teak stand, in order to better understand the ecology of these important plantations. It is envisaged that the timing and movement of nutrients will be related to seasonal climatic and growth patterns. However, it is unclear how closely these coincide or if there are major losses and gains to the system. It is important to clearly understand that the focus is an in-depth understanding of the stand ecology and that each of the various sections are interlinked.

Site

The site should be flat or have a very gentle slope, and not subject to excessive overland water flow, pending of water or erosion. Soils should be typical of the region. A well stocked stand of 20-30 years old should be selected. There needs to be an open area nearby for collection of uncontaminated rain.

Layout of plot

Some 10 temporary plots (with at least 12 trees) should be laid out in the compartment and their stocking, basal area (BA) and mean girth obtained. This information will be valuable when later (say 0.05 -0.1 ha) permanent study plot should be established, containing at least 20 trees, as well as sufficient space for the various studies. The corners of the plot should be pegged and the plot fenced off against stray animals.

All trees in this plot should have their BA level marked, numbered and be measured for girth and total height at the start of the study, when the trees have lost all their leaves and at the end of the study (after 365 days).

The key measurements to be obtained over a one-year period are: Biomass and Nutrients in three occasions (to estimate internal cycle and nutrient accumulation patterns).

Biomass and nutrient patterns

At the start of the study five trees should be sampled for biomass and nutrients when the trees are in full leaf and well before leaf-fall to get samples of the maximum nutrients in the foliage and stand. The biomass relationships from these five trees may be used to predict the biomass and nutrient accumulations of three occasions; at the start of the study; when the trees have lost all their leaves and at the end of the study year. The girth and height measurements on all plot trees are critical for these estimates.

The five trees should be selected to cover the diameter range and should be normal looking (i.e. not malformed or excessively branched etc). The methods used should follow the standard procedure, taking particular care in getting representative samples of foliage, twigs, branches and branch wood. These trees must come from close to the permanent plot but not within it. Nutrient analysis will accompany the biomass study and will be appropriate to use for estimating nutrients at the beginning and end of the study. However, representative branch wood, branch and twig samples need to be collected from five trees when the trees have lost their leaves, and analyzed for nutrient concentrations. The purpose of these samples is to estimate the internal nutrient cycle and the storage of nutrients in the crown. Nutrients are probably also stored in recently formed xylem, the cambium and phloem of the stem and roots; it would be a challenge to estimate these without felling the trees but could be done, if desired.

The modified basal area ratio method may be used to convert data to a per hectare basis.

A Soil nutrient pools

This should be sampled on the three occasions when the measurement of the main study plot, described above, are made.

The sampling scheme should follow the standard procedures used for the chronosequence study. However, in this case a measure of soil variability is needed so the samples should not be bulked in the same way. It is suggested that the plot be divided into 10 x 10m squares and bulked samples be collected from each of these quadrats (5-10 cores/quadrate-standardize this).

Soil descriptions (pedon just outside the plot) plus bulk density are also required. Collect a 0-20 cm bulk density sample from each of the quadrates, plus deeper once from the pedon.

Statistical analysis will give estimates of the means and repeated ANOVA might also be a useful technique.

A Litter fall

This will be sampled inside the plot. Some ten 1 x 1m randomly located subplots (quadrats) should be laid out. These subplots should be numbered so they can be sampled in the same order each time. The forest floor should be cleaned and covered with shade cloth (1.5 x 1.5 m and pegged to the ground) to reduce soil splash. A portable sampling frame of 1 x 1 m should be made. At the time of sampling this should be laid over the litter and material within it carefully collected. Other litter on the outer part of the cloth should be swept aside.

Label samples and bring to the laboratory where the individual dry weights can be obtained. Each sample should be divided into three components; teak leaves, teak reproductive parts and other material (bark, branches, other litter etc). Quadrats 1-5 and 5-10 may be bulked to give a total of six samples for chemical analysis per sampling period.

Normally sampling should be at monthly intervals, but extra could be justified at the time of peak litter fall, particularly if rain is likely. The study should continue for a full year.

Statistical analysis will give estimates of the means and repeated ANOVA might also be tried.

Litter breakdown study

Litter break down studies usually employ litterbags, but these have serious limitations such as not allowing entry of many insects. There is an alternative method that has been described by Anderson and Swift (1983) in which the litter is in direct contact with the soil. This method has not been tested in teak and it is possible that it will be difficult because of soil sticking to the decomposing litter.

The experiment should consist of a split-plot, randomized block design with four replications and 8 treatments. These 8 treatments consist of 4 main plot treatments, which are to study litter collected at quarterly intervals (representing the main climatic periods). The subplots are the two methods-litterbags and non-bag techniques. For each of these combinations decomposing litter will be collected at monthly intervals for up to six months.

Ignoring the latter sampling over 6 months for now, the central statistical design is:

Source	df
Replication	3
Season	3
Error a	9
Methods	1
Season X method	3
Error b	12
Total	31

Note: It is important that the placement of the different seasonal samples be randomized within each replication, and the order of the paired methods be randomized also.

Litterbags should be made of nylon or similar and have a mesh size of 1.5-2 mm. The same mesh will be used to cover the samples inside the metal rings that are in contact with the ground.

Use about 25 g of material; the exact weight needs to be known for each bag but it need not necessarily be the same in all bags provided. Each can be identified in the field. Collect very recently fallen litter (litter can quickly loose some soluble nutrients). In some months the litter will be leaves but at other times other material will predominate. Use the main type of litter falling for that period.

Random samples of the litter put in the bags must be returned to the laboratory to obtain its moisture content and for analysis. Subsequent samples should be collected at monthly intervals. Dry weights should be recorded to nearest 0.1 g.

An Interception, Through Fall and Stem Flow

An Introduction: Dynamics of an ecosystem can be better understood by studying the movement of water through its various components since the nutrients in incident rainfall usually represents an input to the system. Incident rainfall in the forest canopy is dispersed in three ways–

(i) TInterception by canopy
(ii) Through fall .
(iii) Stem flow

Interception refers to that part of the precipitation retained by the aerial part of vegetation, which is either absorbed by it or is evaporated. Forest canopy intercepts the incident rainfall so that its kinetic energy is dissipated by the aerial portion of the trees rather than imparting to the soil and hence, forests are effective agents in reducing soil erosion, flood etc. Except intercepted rainfall, rest of the amount of rainfall reaches the ground surface either by passing through tree crown, i.e. through fall (or) concentrated by the leaves and branches and later channeled down the boles to the forest floor, i.e., stem flow. Different vegetational types greatly influence the water regime of the area and may route precipitation through the basic pathways in different proportions.

Rainfall inputs

Rainfall in the open is important for two reasons. Firstly, it is needed for interception calculations and secondly, for nutrient inputs. It should be stored frozen or close to freezing, prior to analysis of N, P, K, Ca, Mg, and Na.

For this reason the site needs to be reasonably close, yet one where extraneous contaminations are unlikely. The rain gauge should be elevated to avoid soil splash.

Two 10 x 10m representative plots were demarcated on each forest tree species stand. Circumference at breast height (Cbh) should be measured from ten trees representing different girth classes. Canopy cover (% ground surface and area (m2 trees) should be measured through line interception (Misra 1968).

Through fall

To study the through fall the transect technique be used. Three transect collectors, each about 3 m in length and 15 cm in width should be randomly located in the plot. Each should lead to its own large collector that can be used for sampling the water for analysis and for measuring the volume of water. The three collectors will give a measure of variation, so volumes are being collected in each and samples for nutrient analysis need to be collected and recorded individually. Furthermore, the transect collectors may be moved periodically to ensure an even better estimate. Each time the collectors are moved, this should be noted just in case there are large fluctuations that need to be explained. It is envisaged that the through fall will be sampled in two ways. For most of the time samples will be collected at regular intervals. At other more intensive study periods, samples will be collected on a rainfall event basis. It is important that the size of rainfall events varies widely, from light showers to very heavy downpours, as this will assist in regression fitting and data interpretation.

Two types of data will emerge. One will give the annual through fall and include a measure of nutrients in it. The second will look at canopy interception (e.g. the amount of rainfall to wet the canopy) in relation to rainfall intensity. Interpretation of this data would be assisted by a measure of canopy area.

Through fall can also be measured by placing polythene containers beneath the canopy of each tree in a stratified random distribution. Flasks should be moved to different positions after each collection. Cotton wool may be put into the funnel to prevent contamination with organic debris (Liken and Eatran, 1970).

Stem flow technique

This could be measured on five randomly selected trees within the plot. Care must be taken in:

(i) Selecting trees to cover the diameter range so the trees are best selected on a stratified random basis. However, reject very abnormal trees (say with badly broken tops) or where the stem does not allow the collecting tube to be installed.

(ii) Using inert materials for the collecting spiral around the tree.

(iii) Using containers that do not allow light in, otherwise algae etc. will grow.

Again sampling can be a mixture of regular and even based. It should be possible to relate stem flow to tree diameter or basal area, using regression techniques, either on a seasonal, annual (or) size of event basis. The annual information can then be expressed on a land-area basis, based on the diameters of the trees in the main study plot. Confidence intervals or other measures of variation can be placed around these estimates.

Stem flow from the marked trees can also be collected through polythene collars (Liken and Eaton, 1970) into polythene container. For this a spiral polythene gutter of 4 cm diameter should be fitted in each stem with the help of iron nails of size 1 feet (2 nails per cm) and later sealed with paraffin wax. The gutter may be fixed at a height of 1 .0 m above the ground surface. A plastic funnel must be attached to the lower end of the gutter connected to a polythene container. Cotton wool should be placed in the mouth of funnel to prevent the entry of extraneous material received. Stem flow for a tree may be calculated by dividing the stem flow volume by its crown area. Canopy retention can he-calculated by subtracting the net precipitation (stem flow + through fall) from gross precipitation.

Leaching Method

I recommended using tube lysimeters. It in made from small stainless steel tubes about 5 mm in diameter, with a ceramic Tube of about 2-3 cm length attached to their bottom ends. A narrow flexible plastic tube (as might be used in auto analyzers) goes down inside to the bottom of the tube. It is sealed as it passes out to the top of the tube, then goes through a rubber bung into a bottle via a thin metal tube. A vacuum is created in the bottle and this draws water from the soil through the ceramic, and the plastic tube into the bottle. This process may take several hours. The solution is analyzed for nutrient concentrations.

Tube lysimeters could be placed at two or three depths. Those at 20 cm would give a measure of the solution strength the main feeding roots are in. Those low in the soil profile would show when and if significant amounts of nutrients are being lost by leaching. To convert from concentration to a quantitative value of what is lost, then either soil moisture has to be measured down the profile, or reliance placed solely on estimated evapo-transpiration using models such as those by Berryman and Trafford (1974). Another approach may be to try and make a mass balance of a very mobile ion such as chloride. Probably, for the purposes of this study, estimating evapo-transpiration from climatic data is all that is required in the first instance. If very high concentrations are found at depth during some seasons of the year, then more detailed supplementary studies can be made.

For this study it is suggested that five randomly placed replicates be used and three depths studied (20, 50 and 90 cm).

The tubes should be placed in groups of three. This number of replicates is chosen to give a measure of variation and also because sometimes the seal of the ceramic is broken. The tubes arranged in groups of three will be relatively easy to statistically test if nutrient concentrations change with depth. Begin the study at the start of the rainy season as the soil profile is wetting up. Take sample at regular intervals. The samples should be analyzed for NH4+, NO-.. P|. and other actions. Note that it is important to look at the form of N. Conductivity might also be useful if there are sufficient snrnples.

Seasonal Pattern of Tree Growth

Part ot the study envisages relating the nutrient fluxes during the growing season to tree growth. One easy way to do this is to install dendrometer bands on 5 or 6 trees (at breast height) within the plot and measure these at regular intervals. Dendrometer bands are made of thin, non-corrodible strips of metal held in place by a light spring, so that they encircle the tree and fit snugly to the bark of the tree. They are fitted with a vernier scale, so changes as small as 0.1 mm in diameter can be recorded.

Nutrient Movement in Leaves

When the leaves flush there will be a large movement of nutrients from storage within the tree to the new growth. Similarly, there will be a movement of nutrients from the leaves back into the tree before the leaves are shed. This process could be studied in more detail by periodically collecting leaves during the growing season. We proposed that this be done at this site, because it will give better information on the internal cycle and demand for nutrients. It ties in with the biomass study also which gives a measure of how much and where the nutrients are in the tree and also the total amount of mature foliage. As there will be regular work going on in the plot, collecting this information will not add greatly to the workload.

Access is needed to branches in full sunlight. On each sampling date, take three complete leaves from each tree to give 12 leaves per sampling date. These leaves should be about average for that time of the year but be randomly collected. Collect at monthly intervals, or every two weeks, if rapid changes are occurring. The number and dry weight of the leaves, from the separate trees should be measured and recorded and then all 12 combined for analysis of nutrient concentrations. Insect damage to the leaves may occur, particularly late in the growing season. Therefore, it is recommended that an estimate be made on the amount of damage to each leaf. Leaf eating insects are part of the nutrient cycle.

From this data, nutrient contents can be estimated on a per leaf basis, and also, using the tree biomass data, the total movement (on a per hectare basis) calculated. The information will also be useful in defining the best time to sample teak for nutrient diagnostic purpose.

An Other Climate Data

It would be useful to have access to or actual measurements of standard meteorological data. A weather station in the open would be ideal

3

Technique for Nutrient Requirments of Forest Specis

General Introduction

In many countries recent investigations have shown that one of the reasons common for sub-optimal growth in nurseries and plantations is the absence of one or more essential nutrients. Afforestation of various sites in India also has not always led to high production. This may be due to the fact that many of these sites being marginal in character might not have among other things adequate amount of nutrients to sustain good growth. It is on such sites that the knowledge about the nutrient requirement of a species and the use of chemical fertilizers in forestry will be of paramount importance. Numerous trials have shown that in forests, particularly in the plantations, the yield could be increased manifold if proper dose and regime of application of fertilizers could be worked out (George, 1977, Shanmughavel, 1995).

It required to fulfill three objectives in this chapter. In the first instance, an attempt has been made to discuss briefly and critically various methods in the study of forest nutrition. Secondly, to summarize the work done in India particularly in view of the fact that many authors (Nickolas, 1957, George, 1977) have emphasized that very little information of the nutrient requirements of forest plants is available from tropical countries, and thirdly to indicate future course of research.

An Aims in Fertilizing Forests

The following are the main objectives of fertilizer application.

(a) To ensure increase in productivity at economic levels of fertilizers.

(b) To increase and maintain fertility level with a view to sustain high productivity.

(c) To accelerate growth and help establishment of transplants in the early stages against competition from weeds.

(d) To obtain quality seeds in sufficient quantity from seed orchards and plus trees.

(e) To improve vigour and resistance of multipurpose trees.

This is essential to have preliminary information which will elucidate the following five basic principles:

(a) What are the limiting nutrient requirements of the species to be used?

(b) What capacity do the existing soil types have for supplying these requirements, both from the viewpoint of reserves and rates of release?

(c) What course will soil fertilizer reactions take when fertilizers are applied and at what rate will these reactions proceed?

(d) What side effects will result from application of fertilizer at various rates?

(e) What chance is there that treatments can be applied efficiently under a given site condition?

A clear understanding of the above questions requires a thorough knowledge about the behavior of the species, soil types, fertilizers and even other conditions prevailing on a particular site. Many methods, each one of them having its own advantages and disadvantages, have been adopted in solving these questions which are briefly discussed under the following heads.

DIFFERENT APPROACHES ADOPTED IN THE FOREST NUTRITION RESEARCH

1. Visual symptoms.
2. Pot analysis.
3. Soil analysis.
4. Tissue analysis (foliar diagnosis).
5. Indicator plant methods.
6. Substitution methods.
7. Field trials.

Diagnosis of visual deficiency symptoms

(a) By Pot Culture:

The most preliminary and rapid method to determine a nutrient deficiency is by observation of visual symptoms. The standard procedure consists of growing plants in sand or water cultures (sand cultures being preferable to water cultures as they provide a solid rooting medium and also provide better aeration) under controlled conditions and fed with solutions of pure chemicals. Deficiencies of the individual nutrient elements are created in the nutrient solution by omitting each separately from the complete nutrients solution and the symptoms of each deficiency exhibited by the plant are recorded (Wallace, 1951).

This method is especially useful in the assessment of effects of various treatments (fertilizing, thinning, mulching, etc.) where other measurements fail for some reason.

Difficulties and inaccuracy in the interpretation of results obtained in such experiments under field conditions arise when a plant suffers from lack of more than one nutrient or when a deficiency of one nutrient combines with an excess of another, or even when the symptoms are modified by the environment (Vanclay, 1992).

However, this is essentially a qualitative approach for nutrition studies. It is less reliable in the case of mild deficiencies, i.e. in the range of so called "hidden hunger" in which growth may be substantially depressed despite the absence of recognizable "distress signals". Thus visual symptoms may be frequently valuable indicators of nutritional stress, but the symptomatology is, at best, a subjective and inexact science. Such studies need to be followed by factorial experiments where various levels of nutrients are tried under controlled conditions reflecting nutrient interactions.

(b) By Foliar Spray:

In recent years, it has become increasingly popular particularly in fruit trees to apply nutrients as foliar sprays. The usefulness of foliar application of nutrients depends on the following circumstances: (a) the existence of special problems that may not be coped with as well by application of fertilizers to the soil or by soil amendments (b) satisfactory response to the nutrient sprays, (c) economic materials and methods of application. Because of fixation problems, micronutrients e.g. Fe, Zn, Mn and Cu are usually applied as foliar sprays. This method is also useful where a temporary control is needed or response to soil treatment is slow or to correct the balance of nutrients when it is not possible through soil dressing (Bould and Hewitt, 1963).

By Pot trials

The time scale is strongly against the experimenter. In forestry it is presumed that even with rapidly growing species, it will take about seven years to discover from field trials alone whether the treatments in an experiment are adequate as to the rate and nutrient tested. Commercial evaluation will take as long again. Therefore, prior to beginning the field trials, many workers have adopted a more physiological approach, viz., experimenting with seedling or transplants in pot or solution culture in order to gain basic information about the nutritional behavior of a species in comparatively shorter period. The special value of pot cultures depends more on the fact that they can eliminate and then partly reintroduce under controlled conditions, selected combinations of the complex of factors comprising a soil environment. Cultural methods permit the study of interactions of deficiencies and excesses at accurately determined levels (Hewitt, 1970). Some of the obvious advantages of these methods are (i) more complicated experimental designs can be used without prohibitive cost, (ii) variation in growth due to other sources than nutrients including those of genetic origin can be reduced to the minimum, (iii) there are greater possibilities of evaluating total growth response in more details and more correctly than in forest stands, (iv) most of the possible indirect effects of fertilizers can be entirely eliminated in pot cultures, (v) in sand and water cultures, nutrient interactions noted or suspected in the field can be tested (Gentle and Humphrays 1968).

A mainly three kinds of pot cultures being used for studying nutritional behaviour of species, viz., (i) sand culture, (ii) water culture and (iii) soil culture. Sand cultures are simpler to manage and provide more satisfactory growth medium for most plants. They have the solid rooting medium with better aeration and permit greater latitude in concentration of toxic elements. Phenomena like root contract and ion exchange can be studied more easily in sand cultures. Water cultures on the other hand, benefit more conveniently the change of nutrient solutions at specified times and roots can be thoroughly rinsed if desired. Contamination by other nutrients can easily be avoided. But water cultures require special efforts for aeration and to check algal growth on the walls of the containers. Soil cultures or the so called soil bio-assays have been used with good effect to make direct biological measurements of the nutrient supplying powers of forest soils. Basically the technique consists of growing tree seedlings in sample of soils to be tested, and measuring the response obtained when nutrients are added singly and in all desired combinations. Pot cultures have also been used to study relative merits of different culture solutions, forest hardiness, drought resistance and even potential water requirements of forest species.

Most of the disadvantages of such cultural experiments which reduce their usefulness to a great extent are, (i) differences in plant growth observed in different media viz., sand, solution, gravel, mulch, compost or soil have not been established and hence results of these experiments in different media are not comparable; (ii) in sand cultures wetting and drying cycles are too artificial unless very expensive equipments are used. Root cat ion exchange action is not complemented with the natural soil action exchange mechanism of maintaining buffering action which is inherent during clay and humus portion of most natural soils; (iii) algal growth in containers and bacterial infection of roots in solution culture are difficult to control under some pH and nutrient regimes; (iv) pot trials using the soil which may be ecologically very different from the same amount of soil *in situ* are of limited value because of the change in physical properties due to being a very limited part of the profile. Even the monolith soil samples are disturbed to some extent. Subsequent drainage and drying cycle conditions cannot be compared with those of the natural soil, nor can soil temperature regimes be simulated in pots. Even the chemically important changes in the availability can occur due to differences in microorganism population resulting in depression or stimulation of growth. K-deficiency, for instance, has been noticed in pots at early stages whereas field experience reveals no such deficiency in these cases; (v) the nutrient uptake and requirements of seedlings may be different from that of adult trees because of differences in anatomy and cell physiology. For example, visual symptoms of K-deficiency differ considerably between seedlings and adult trees. It has been noted that soil bioassays for reasons, which are not fully understood, tend to exaggerate the response to addition of phosphorus. Kauland Mathur (1971) found differences in branching habit by the deficiency of N and P in seedlings *of Eucalyptus sp.* which may not occur under natural conditions in adult trees; and (vi) the results are not immediately applicable to the field conditions because no way has been found to relate absolute levels of nutrients used in pot or solution cultures to those found in field soil. Field trials (and in some cases, supplemented by other methods) on the basis of these experiments are necessary to get conclusive results.

In sum up their usefulness for practical purposes, it can be said that various kinds of pot trials, in spite of the above mentioned limitations and need for improvement in techniques have an important part to play in providing excellent information on nutrient physiology of tree seedlings.

Soil analysis

In the study of nutrient requirements the value of soil analysis of forest species has to be considered separately for tree seedlings in nursery beds and fertilizer needs of forest plantations and second growth stands. During the past three decades, foresters in advanced countries have made remarkable progress in the soil requirements of nursery soils. Avoidance of excessive stoniness and a silt plus clay content between 15 and 20 percent obviate difficulties of cultivating or transplanting and assume reasonable water retention. Impermeable or calcareous substrata less than 1.2 m from surface should be absent. The top 15 cm should contain the appropriate mycorhiza, around 2% organic matter, 0.07 per cent total nitrogen, 10 kg/ha of available phosphorous, 80 kg/ha of available potassium and 1.5 and 0.5 meq /100 g respectively of exchangeable calcium and magnesium. Soil pH should be between 5 and 6. It is well known that nursery stock removes appreciable quantities of plant nutrients, which have to be replenished in routine fertilizer treatments.

In the study of nutrient requirements of plantations and forest stands by means of soil analysis presents some formidable problems and has not proved very reliable partly because the investigators have tried to establish some correlation between the nutrients and soil biomass components. Moreover, very often one or few of several soil factors; both physical and chemical have been picked up for correlation studies, which is risky. However, multiple correlation studies by electronic computers may greatly facilitate provided correct parameters are used.

George (1977) recommends prior analysis of soils in planting sites to ascertain the presence of an adequate supply of essential nutrients. Such analyses are particularly desirable in regions showing poor growth of native stands, or plantations. As a rule, the analysis could be limited to the upper 6-inch layer of soils as its composition usually reflects the properties of the substratum and is of decisive importance in the initial growth of planted trees.

May recent works have emphasized the importance of parent rock in the study of nutrient requirements of forest trees. Forest trees have a long span of life and their roots are capable of extracting more nutrients from soil minerals than are those of agriculture crops not so much because of higher efficiency as of longer persistence.

The important which have a great bearing on the turn over and nutrient availability in the soil and ultimately on the value of soil analysis as a guide to fertilizer needs of the forest crops are– (i) the complex nature of humus layer, (ii) availability of moisture, (iii) aeration and (iv) biological activity of soil organisms. The availability of nutrients after the application of fertilizers depends to a great extent on the situation of these factors

1. Soil sampling technique

As a result of differentiation of distinct horizon, often differing appreciably in composition, most forest soils are markedly heterogeneous. On the other hand, most trees are long lived, deep rooted and exploit large mass of soil. Therefore, there is a controversy about the horizon of the profile, which should be studied in detail. These factors introduce considerable soil sampling problems particularly in tropics where root penetration may extend over several meters. In practice, sampling is done up to 3 m depth

2. By Method of extracting nutrient

Soil samples can be analyzed in many ways, viz., HCI soluble nutrients, exchangeable basis, and determination of total exchangeable ions, available nutrients or measurements of ions absorbed by ion exchange resins mixed with soil to determine their nutrient contents. It is not yet definite which of the above method will give a true picture because of the dynamic character of the available nutrients in soils. For instance, change in the amount of nitrate nitrogen in soil may be caused by the change in the rate of decomposition of nitrate or ammonia.

3. By Growth factor

The dynamic nature of growth of the plant itself introduces problem for predicting the fertilizer needs. An intensely growing forest stand may exhaust the soil supplies of easily soluble nutrients before growth ceases resulting in a negative correlation between yield and soil nutrient content.

By Tissue analysis Foliar diagnosis

In this method of plant analysis is based on the concept that plant itself is the best indicator of soil nutrient availability and physiological uptake and the analysis of the plant tissues constitute a natural biological rather than an artificial extraction (soil analysis) method reflecting integration of numerous inter-related factors affecting the availability of nutrients.

1. By Sampling

A Strict adherence to sampling standards is a pre-requisite in foliar analysis particularly in case of trees having long span of life and usually a large well developed crown. The important aspects in sampling procedure are (i) age of leaves and number of trees, (ii) season, (iii) position in crown, (iv) shading, (v) fruiting, (vi) weather conditions, (vii) time of day and (viii) attack of parasites. Adhering dust particles and pollution in industrial areas may also act as a source of error. This indicates the need for standardization of sampling procedures. According to most investigators conifers should be sampled at the end of the growing period and the deciduous trees just before autumnal yellowing. Regarding the position in the crown it is agreed to sample as near the tip as possible, and in conifers young needles from a particular whorl.

2. On Year to year variation in nutrient content of foliage

Generally there is appreciable difference in the level of certain elements in different years. Srivastava et al., (1972) found large variation in litter fall and composition from year to year (The factors responsible for the variation are rain and drought, leaching from leaves and even intensity of transpiration stream. For definite conclusion it is therefore advisable to sample leaves for more than one year.

3. By Handling and preparation of samples

Rapid killing and drying after collection is necessary to avoid decomposition changes and apparent increase in dry weight and mineral content by continued respiration.

4. By Unit of expression

General practice is to express macro-elements in terms of percentage on dry weight basis and microelements in ppm (parts per million). Several authors (George, 1977, Shanmughavel, 1995), have used ratios between the various elements (percentage per 1 00 mg) instead of or together with dry weight percentage, some of which viz., K/Ca, K/ Mg, Mg/ Ca may be the expressions of antagonism between the cat ions in question. Mengal and Kirkby (1987) by using a standard punch of one cm^2 expressed results on leaf area basis and Kessler (1961) based their results on the amount of nutrients per 100 needles.

5. By Interpretation

Interpretation of data obtained from leaf analysis is of decisive importance and presents some difficulties because of numerous interactions, ion antagonism and differential uptake which are a rule rather than an exception. Further, the amount of a nutrient in the foliage at any time depends not only on the availability in the total mass of soil accessible to root system but also on the factors that influence the growth of the plant. Therefore, knowledge about the dynamic nature of growth and nutrient, their interaction and movement in plants, specific functions, varying requirements of plants in different seasons and ages together with the role of environmental factors is essential for correct interpretation of the data. A fair amount of knowledge has. been accumulated on these but much still remains to be done and indicate a pressing need for intensive basic research.

In many cases, nutrient interactions coupled by growth are so complicated that simple correlations are not possible. If growth increases more than the nutrient uptake in a vigorously growing stand, the result will be lower concentrations (a dilution effect or "false antagonism"). This is particularly observed when supply of the same nutrient is increased from extreme scarcity to optimum. On the other extreme there is often a tendency for luxury consumption, i.e., accumulation of nutrients beyond the optimum level without corresponding growth (Leyton and Armson 1956).

In the event of multi- elemental deficiencies, a number of workers have analyzed the relationship between growth and concentration of different nutrients in the foliage in terms of multiple regressions, which has been facilitated by the modern computers. This approach implies that height or site index is a function of nutrient status and factors making significant contribution to the regression of height on foliage composition can be distinguished from factors related to growth by virtue of their internal relationship with the significant factors.

Another approach for development of diagnostic issue analysis may be found in fractioning the organic and inorganic compounds in plant part. It is a well known fact that a lack of plant nutrients may cause disturbances in metabolism. Some of these effects may be specific enough for diagnostic purposes (Kessler, 1961).

In the light of the above discussion, it is evident that the methods of foliar diagnosis if used with due care in sampling and interpretation will be highly useful in determining nutritional imbalance in nursery stock as well as plantations and forest stands.

By Indicator Plant method

1. By phytosociological approach

The method is based on the concept that each species has got certain ecological amplitude, i.e., it can thrive (as vigorous specimen) under certain environmental conditions and hence is a measure of those conditions. Thus it is based on the cause-effect relationship where the effect is taken as a sign of cause (Srivastava et al., 1972). In literature many examples can be found where indicator species have been sorted out and certain 'correlations have been established with the growth and productivity of the dominant species (Cajender, 1926), with the occurrences of natural regeneration (Switzer et al 1972) and sites suitable for afforestation with exotic species (Havel 1970). According to George (1977) for a practical use in silviculture, it is not necessary to know exactly about the demand of the indicator plant: it is sufficient if the sites can be classified into units with different properties concerning actual or potential use in under forest management with the help of indicator plants.

Yet there is another type of indictor, so called "nitrate plants" a term used by Goodwin (1954) for plants demanding nitrate in soil and accumulating in leaves to become established and thrive. It is also possible, sometimes to use microorganisms as indicators of the soil supply of nutrients. Aspergillus has been used extensively for Mo determinations in arable soils. However, there are no such examples on forest soils.

2. By Use of indicator plants for extraction of nutrients from the soil

In this method, indicator plant is grown in pots or in the field to extract nutrients to ascertain the amount of available nutrients. The most widely known pot test method is that introduced by Freez, 1961 in which red rye seedlings is determined after 17 days. The uptake from the soil is then calculated. Tables are available to show the quantities of nutrients that should be available for food growth of various crops. A similar method with tree seedlings as test plant has been used extensively in forest soils (Hibberred. 1991). However, as in the case of soil analysis and pot culture using soils, in this case also, it is doubtful whether the method gives a correct picture of the amounts of nutrients available in forest soil *in situ*.

Substitution method

In this method, an attempt is made to determine the total nutrient consumption by a forest stand by analyzing harvested material as well as litter. This kind of work was the first serious attempt to obtain information on forest nutrition. Thus George (1977), studied the amounts of nutrient content in plant parts and explained site deterioration after centuries of litter removal. This approach gained immense importance after the article by Rennie (1955) and after that many workers (George, 1977, Shanmughavel, 1995), have studied the mineral cycling making a complete balance sheet for nutrient supplies and losses in different ecosystems.

By Field trials technique

A field trials are the traditional method of testing fertilizer responses. The data can often be used to determine precise quantitative relationship between the levels of nutrients and yields. Factorial experiments can be useful in studying the interactions between mineral nutrients and deficiencies resulting from lack of balance between elements. Field experiments, however, have to be used at final stage of work. According to Gentle and Humphreys (1968), if used for primary diagnosis, they run the risk of entirely missing the crucial deficiency or failing to differentiate responses between the essential and non-essential elements in the fertilizer. They may indicate nothing about the mechanism of response or why response continues or fades out. In order, therefore, to avoid wastage in time by using "hit or miss" technique, the experimenter needs to collect all available information on the nutritional behaviour of the species and the soil type where the species is to be tried prior to the beginning of the field trials.

In addition, a few practical criteria merit particular attention when field experiments are designed (Gentle and Humphreys 1968):

1. The object of the experiment should be clear and as simple as possible because of the necessity of keeping down the total area of the experiment to a reasonably uniform area (as in the forest it is often difficult to find large uniform areas).
2. The object should be pertinent to the subject. It is, for example, of little value to experiment with K and P in a stand showing nitrogen deficiency. Many fertilizer treatments proved inefficient because the applied fertilizers did not include the elements necessary for the correction of existing nutrient deficiencies.
3. The experiment should conform as far as practicable to future forestry practice in respect of spacing, site preparation etc,
4. The design should be simple enough but statistically sound. Thus the plots should be randomly assigned at least in duplicate for each treatment and combinations.
5. The plot should be spacious enough to carry through at least two thinnings so that silvicultural interactions are tested. Beside the above, it should be ascertained as early as possible that a certain treatment has desired effect and no important side effect. Because, the indirect effect of fertilizer treatments are even more difficult to interpret. To avoid this difficulty, the effect of treatment can be tested by treating plants with some other means, say by spraying or stem injection in case soil application of fertilizers is ineffective or has profound side effects.

There are numerous reports in which application of organic matter has increased forest growth. Instances have also been reported in which cultural operations like soil working, weeding and trenching have improved growth of young trees. Such treatments should not be used while investigating nutrient requirements with the help of field trials. Their effects are too complex, affecting micro-biological conditions as well as soil chemical and physical properties. It is then difficult to distinguish between the effects of fertilizers and such treatments.

One more precaution necessary in field trials is to transplant the seedlings as naked plants (though it involves the risk of mortality of a few seedlings) and not along with pot soils, as has been the practice lately since the adoption of polythene bags as pots. In the later cases, the plant roots remain confined for a year or two to the pot soil, which is well prepared and rich in nutrients as compared to the soil on planting site. Therefore, the fertilizers applied during this period go a waste, either due to leaching or utilized by the ground floor vegetation.

It should also be noted that some commercial fertilizers contain magnesium, sulphur or trace elements in addition to the major nutrients. Therefore, special attention must be given to the choice of the fertilizers if the effect of the various nutrient elements is to be assessed correctly. A positive reaction to an added nutrient by the sample trees can usually be taken as evidence for a deficiency of that nutrient. This procedure is expected to indicate within 5-15 years whether or not the procedures used are likely to be applicable on large scale forestry, and if so whether biological and economic objectives are likely to be achieved. The more closely experimental design confirms to pilot plot operation, the more expeditiously can useful information be fed into the normal operations. As a rule, a year spent in -this way is worth about 5 years spent filling in data deficiencies later in the life of the field experiment (Gentle and Humphrays 1968).

Field trials with fertilizers were initiated during fifties of the past century in European countries. In most trials positive results were obtained with the use of green manures particularly lupin and with the application of organic manures, e.g., litter and slash. A considerable improvement in the growth of plantations due to application of lime and crushed rocks rich in bases was observed. The effect of soluble mineral fertilizers was inconsistent and often short lived (Wider and Lang, 1982). Wood (1960) attributed conflicting results of early fertilizer trial to the following causes: (i) insufficient knowledge of soil chemistry, action of fertilizers and nutrient requirements of tree species, (ii) competition of weed vegetation which developed vigorously on the fertilized soil, and (iii) occasional drought or winter kills which not only annulled the benefit of the fertilizer treatment but made such treatment harmful.

However, the scheme is so large and the climato-edaphic, physiographic and even biotic conditions will vary so much besides the personal errors caused due to so many separate terms looking after the experiments that it will be extremely difficult if not impossible to sort out the effect of fertilizers. The design of the experiments also appears to be large so as to make it difficult to handle the experiments or even to find out uniform area for laying out experiments. Moreover, it appears that no attempt has been made to find out either the nutrient requirements of the species (the three species may vary) or nutrient deficiency symptoms in plants before deciding upon the levels.

The better alternative would have been to carry out field trials by 4 teams exclusively appointed for the purpose in 4 regions (North, South, Central and Eastern) on 2-3 varying soil and climatic situations in each region after a systematic analysis of soils determining the possible deficient nutrients and keeping the experiments as small as possible, may be in a single or double nutrient combinations. Experiments should have simultaneously been laid out to determine the method and time of application, frequency and source of fertilizers and its economic aspects. Along with soil analysis, foliar diagnosis method should have been applied to find out the response to applied nutrients. These teams if necessary would have also conducted pot cultures to study any special problems arising out of their field experience regarding antagonistic, synergistic or other interaction effect.

BYCOMBINATION OF TWO OR MORE METHOD

Since no single method can give complete and reliable information on the nutrient requirements of forest species, in practice, several possible combinations of two or more methods have been used. Pot cultures with seedling have been combined with observations of deficiencies and toxic symptoms. This was further combined with the chemical analysis of seedling. This approach has been very useful in nursery work and many observations on visual symptoms, plant nutrient uptake and growth were found to be true over a wide range of conditions including mature trees in forest stands (Wallace, 1951).

The method of foliar diagnosis combined with the field experiments and often supplemented with pot and solution culture experimental data will be useful in indicating and double checking as to which nutrient should be augmented. This combination was used by Morey et al (1985) to differentiate a number of species as "Nitrogen demanding intermediate" and "Nitrogen deficiency tolerant". Several later workers followed this approach and tried to standardize the foliar analysis data with the help of field experiments with or without pot cultures (Tamrn, 1960, Wallace, 1951). The use of exponential curve has not been common; instead there is a trend to use multiple correlation method in the interpretation of experimental data (Wallace. 1951, Ley ton and Armson 1956).

It appears that the method of foliar diagnosis combined with experiments in the field, or both in the field and under controlled conditions, forms a very useful approach in forest nutrition research. In fact Tamm (1960, 1964) calls it, a combination of extensive field experiments with foliar analysis as the best approach to the study of ecological nutrient requirements of various species and Gentle and Humphreys (1968) feel that foliar crop logging procedures will be a regular feature in near future to make certain that applied nutrients are being efficiently absorbed and to control rates of release from various sources more cheaply and accurately than would be possible from soil analysis.

With the field experiment and foliar analysis together, soil analysis data can be combined usefully to determine the fertilizer source, method of application and on which soil type to change to another source. Phosphorus, for example is very mobile within the tree so that a few tree roots contacting the bond are adequate to supply it to the tree. Calcium mobility within the tree is at the other extreme so that bonding is not correct solution, where calcium and phosphorus deficiencies are being remedied together (Gentle and Humphrays, 1968).

A Need for Intensive Research on Forest Nutrition

By using fertilizers in forestry is well established in many parts of the world and there is clear evidence that the rate of wood production can be significantly increased on many sites by the application of deficient nutrients. There are classical examples of dramatic responses to phosphate and even certain micronutrients in parts of Australia and New Zealand (Williamson 1983). The duration of the effect of applied fertilizers varies from 5 to 40 years, depending chiefly on the type and quantity of fertilizers applied and the nature of the soil. Climate and drainage pattern also add to the effectiveness with which the nutrient cycle functions. It is also realized that the species at the lower limit of nutritional adequacy are unduly susceptible to disease and insect attack. Moreover, the wood produced by fertilized trees is little, if at all, inferior to that produced by unfertilized trees (Swan, 1967).

On most difficult problem, not yet solved satisfactorily is to determine the capacity of different soils to transmit the effects of fertilizers to the growing stock. This capacity depends upon several conditions including the supply of soil, water, air and nutrients, climate and length of growing season. According to Willey (1985), the concept of fertilizer responsiveness of soils should be placed in the foreground of all forestry fertilization work and deserves elaboration. At the same time, it is essential to study the effect of lack of essential nutrients on the anatomical structure both shoot & root particularly the growing points. These are the most sensitive parts and the first to be influenced by the deficiency of a nutrient. Another field, which requires attention, is to study the effect of lack or application of fertilizer on root regeneration potential. The roots are usually the first organs of a plant to respond to the environmental changes and success of any species depends largely on the adaptability of roots to these changes.

In the tropical countries in general and India in particular are concerned, it is evident that very little information on nutrient requirements of forest trees is available. This is partly due to the fact that in tropics, a tendency is prevailing where the climate and physical properties of soil are the most important factors for forest growth. Although, it may be true in case of well-stocked, scientifically managed natural and virgin forest stands, the same cannot be said about the man-made forest ecosystems, many of which are being established on marginal soils but with the hope of quick and best returns on sustained basis. During the last decade or so, stray attempts have been made to study some aspects at the Forest Research Institute through pot cultures and foliar analysis but with little follow up action in the field. However, sincere and well thought out attempts are needed on intensive basis to develop efficient methods to suit diverse situations on the commercial use of fertilizers to bring research data to operational feasibility. By the time this is achieved, it is hoped that the country will be self-reliant in fertilizer production and quantities required in forestry operations will be easily available. Unless the afforestation program is ready to provide soils of reasonable fertility, fast growing species and even the improved varieties resulting from research in tree breeding will not give anticipated yield and the areas of the struggling non merchantable plantations will only serve as breeding center for parasites. A research of forest nutrition is urgently required both in the context of production and conservational functions of man-made forest ecosystems the service of humanity in the interest of harnessing the environment

4

Technique Used in Chemical Analysis of Wood

General Introduction

Chemical analysis of wood intended to provide methodology in the proximate chemical analysis, pulping and paper making characteristics of the soft and hardwoods. The characteristics may vary somewhat with growth conditions and age of the trees of a specific plantation.

Proximate Chemical Analysis

By Analysis of Wood

In order to be clear (before the description of the method) a general scheme of the protocol is given in Fig 4. 1 with modifications (Alaudin et al 1971, Bhat and Guha 1952, Bhat et al 1955).

Scheme of the Procedure

Equipment required

(i) Electrical saw

(ii) Sieve having 65 meshes/sq. cm. and 200 meshes/sq. cm.

(iii) Soxhlet extractors

(iv) Filtering crucibles with fritted disk of coarse porosity G2 and G3

(v) "Kipps" apparatus using NH_4C1 and H,SO_4 for the production of gaseous HCI flow.

Reagents required

a) Benzol
b) Ethyl alcohol at 95 °C
c)N aOH solution 5% by weight
d) Chlorine gas

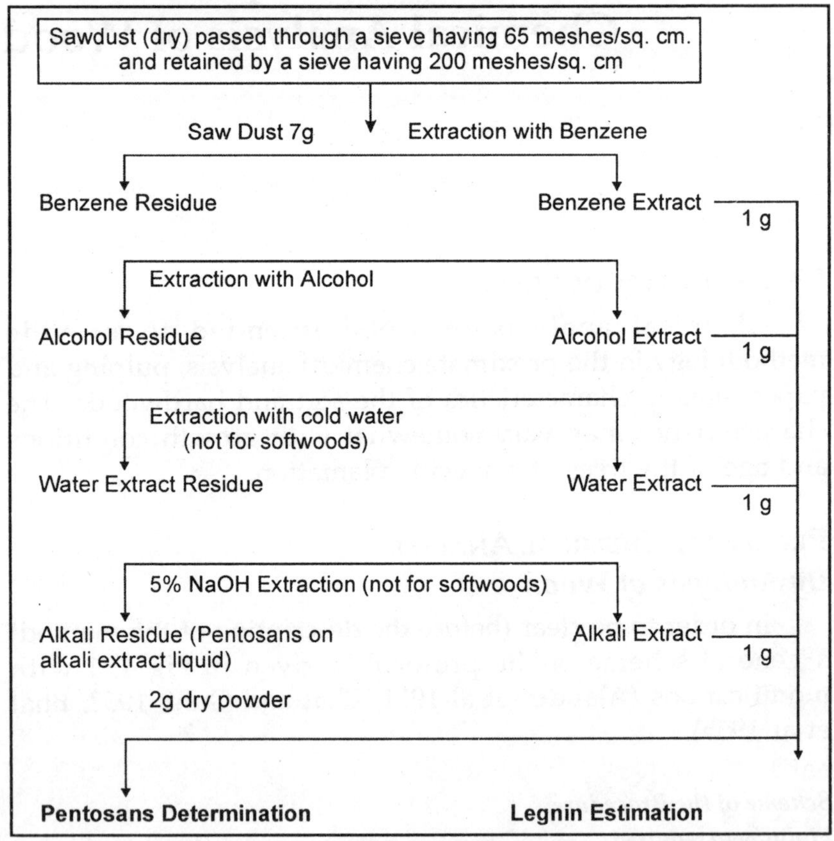

Diogram D.1

SUMMARY: For softwoods the determination of CROSS-BEVAN cellulose is made after the first two benzene and alcohol extractions (Alaudin et al., 1971). Each determination should be made twice.

A Preparation of the wood sample

The wood is reduced into sawdust by means of a saw. The sawdust is sieved on a sieve having 65 meshes per sq. cm. and 200 meshes per sq. cm. The material retained by the second sieve is called "Average Wood" and is used for the chemical analysis.

Benzene extraction Technique

7 g of sample of average wood may be dried in an oven at 100 C. The dried content needs to be extracted with benzene for 6 hours in Soxhlet; the extract should be dried at 100 °C and weighed.

$$\text{Benzene extract \%} = \frac{\text{Extract weight } X\, 100}{\text{Bone} - \text{dry wood weight}}$$

Alcohol extract method

The residue from the previous operation (Benzene extraction) needs to be extracted for 6 hours with 95% alcohol. The extract may be dried at 100 °C and should be weighed.

$$\text{Alcohol extract \%} = \frac{\text{Extract weight } X\, 100}{\text{Bone} - \text{dry wood weight}}$$

Soluble matter in cold water mehod

The residue from the alcohol extraction needs to be dried at 100 C and should be digested for 24 hours with 400 ml of cold water. It may be then filtered through a filtering crucible with fritted disk of coarse porosity G3 and washed and dried at 100°C for 10-12 hours and should be weighed.

$$\text{Cold water extract \%} = \frac{\text{Extract weight } X\, 100}{\text{Bone} - \text{dry wood weight}}$$

Soluble matter in 5% NaOH solution

The residue from the soluble matter in cold water may be treated with 100 ml. of 5% by weight NaOH solution for 24 hours, and the solution should be filtered through the same filtering crucible as used before. The residue needs to be washed with 100 ml of 10% (v/v) acetic acid solution and then with distilled H_2O. The residue at neutral reaction of the filtrate may be dried as before and should be weighed.

$$Alkali\ extract\ \% = \frac{Extract\ weight\ X\ 100}{Bone-dry\ wood\ weight}$$

Cross Bevan Cellulose (with pentosans) method

Two grams of dry material, which had undergone above 4 extractions, should be transferred to a filtering crucible with fritted disk of coarse porosity G3. The filtering crucible may be connected to a vacuum flask immersed in a proper ice cooling container.

It needs to be washed with warm distilled water, and after having put the ice into the container, a chlorine gas flow under slight vacuum should be passed through the filtering crucible for 5 minutes (velocity of the flow: 2 bubbles per second). Then the sample needs to be washed 2+3 times with a 1% by vol. SO_2 solution (about 100 ml). The vacuum may be then disconnected and the fritted crucible filled up with 1% soda solution (the material changes colour to brown). It should be digested with soda for 5 minutes. The vacuum should be connected and the chlorine flows again through the filter. This double treatment needs to be repeated until the material contacting the soda remains white. The solid matter should be dried and weighed. Pentosans can be determined on the residue. Cross - Bevan cellulose:

$$Wood\ theoretic\ weight = \frac{Wood\ weight\ after\ the\ 4\ a.m.extraction\ X\ 100}{100-(sum\ \%\ of\ 4\ extractions)}$$

$$Bevan\ cellulose = \frac{C.B.residue\ weight\ X\ 100}{Wood\ theoretic\ weight}$$

Lignin technique

From the above four extracts, samples of 1 g from each fraction may be placed in four single test tubes and should be treated with 12 ml, of concentrated HC1. Each test tube is closed with a rubber stopper, having an inlet tube going down almost to the bottom and an outlet tube. Through the inlet tube a flow of HC1 gas should be passed which may be supplied by a Kipps apparatus with NH_4CI and H_2SO_4. After this stage the test tubes may be left closed during 24 Hours, the contents then mixed with water and filtered off, and washed thoroughly until the acid reaction of the filtrate disappears. Later the sample may be dried at 100°C for 8-10 hours and weighed.

$$Legnin\,(\%) = \frac{Extract\ weight\ X\,100}{Wood\ theoretic\ weight}$$

Ash

It is determined on 2 g of wood.

$$Ash\,(\%) = \frac{Ash\ weight\ X\,100}{Dry\ content\ of\ Wood\ weight\ 2}$$

Moisture

It is determined on 2 g of wood.

$$Moisture\,(\%) = \frac{Dry\ wood\ weight\ X\,100}{Moist\ wood\ weight\,10}$$

DETERMINATION OF PENTOSANS

Equipment required

1. No rmal distilling apparatus equipped with graduated funnel for filling HCI into the distilling flask.
2. Filtering crucible with fritted disk of coarse porosity 11 G3 (gamma) or 11 G4 (delta).

Reagents required

a) HCI solution 12% by volume (D = 1.06).

b) Hydrochloric phloroglucin solution free from resorcin prepared by dissolving 74 g of phloroglucin in 12% by vol. HCI, then diluted to 10 liters with 12% HCI. The possible presence of resorcin may be tested as follows: a small quantity of phloroglucin is dissolved in few drops of acetic anhydride and brought to boiling; at this moment few drops of concentrated sulphuric acid are added: a violet color indicates the presence of resorcin.

c) Aniline is prepared by adding acetic acid or hydrochloric acid drop by drop to a mixture of equal volumes of aniline and water until a clear solution is obtained.

Preparation of the sample for analysis

One gram of bone-dry wood sawdust may be weighed and should be placed into the distilling flask. 100 ml of 12 % HCI should be added through the proper funnel and should be brought to boiling. After having collected 30 ml of distillate, 30 ml of 12 % HCI needs to be added in the distilling flask, and thus proceeding until 360 ml of distillate may be collected, this amount being certainly sufficient to have all the formed furfural passed into the distillate. (Generally, all the furfural is contained in the first 300 ml of distillate; however until furfural is present, a drop of distillate will give a pink color when coming into contact on filter paper with a drop of aniline reagent). 40 ml of the phloroglucin solution should be added to the distillate, then the volume of the liquid needs to be made up to 400 ml with 12% HCI. It may soon turn greenish-black and furfural-phloroglucide precipitates.

After 16 hours of decanting it should be tested whether all furfural has precipitated, by adding a drop of aniline: if pink color is obtained (presence of furfural) more phloroglucide must be added. The precipitate should be filtered through a filtering crucible with fritted disk of coarse porosity 11 G3 (or 11 G4). Care being taken that the filter may be always filled with liquid. The precipitate should be washed with 150 ml of distilled water, needs to be dried in oven at 100-150 °C for 6 hours, placed into a desiccator until it is cooled and then weighed.

A Calculation of the results

The content of pentosans in the .sample may be calculated from the weight of furfural - phloroglucide (a) on the basis of the following formula:

When the weight of phloroglucide is under 0.03 g:

Furfural = (a + 0.0052 + 0.0020) x 0.5170

Pentoses = (a + 0.0052 + 0.0020) x 1.0170

Pentosans = (a + 0.0052 + 0.0020) x 0.8949

If phloroglucide weight is between 0.03 g and 0.30 g:

Furfural = (a - 0.0052 + 0.0020) x 0.5185

Pentoses = (a + 0.0052 + 0.0020) x 1.0075

Pentosans = (a + 0.0052 + 0.0020) x 0.8866

Over 0.3 g

Furfural = (a + 0.0052 + 0.0020) x 0.5180

Pentoses = (a + 0.0052 + 0.0020) x 1.0026

Pentosans = (a + 0.0052 + 0.0020) x 0.8824

The content in percentage may be figured from the pentosans content in the sample, by the following formula:

$$(\%)\ pentosans = \frac{Weight\ of\ furfural\ phloroglucide + 0.0072\ X\ 100\ X\ 0.8866}{Weight\ of\ bone\ dry\ wood}$$

NOTE: The value of pentosans may also be expressed as content of furfural (furfural deriving from hydrolysis). In such a case the possible errors deriving from presence of methyl-pentosans are lowered.

Methodology to Study the Fiber and Pulp Characters of Wood

During the last few decades, bamboo has become a major source of raw material for the Indian Pulp and Paper industry. The industry is presently facing problems of non-availability of suitable fibrous raw materials in adequate quantity to meet its requirements. The supply of traditional bamboo species for pulp making has been dwindling in recent years. Still bamboo accounts for sixty percent of the fibrous raw material requirements of the industry. With the gradual shortage of such raw materials, the pulp industries are forced to look for alternative stock (Shanmughaval and Francis 1996). Hence a methodology to study the fiber and pulp characters of the soft and hard woods is described herein.

Sample Preparation for analysis

The wood to be analyzed should be reduced by means of an electrical saw. The sawdust needs to be sieved sequentially through a 65 mesh and a 200-mesh screen. The material retained during the second sieving is "average wood" and should be used for the chemical analysis.

Bamboo *(Bambusa bambos, Dendrocalamus strictus),* Eucalyptus *(Eucalyptus tereticornis),* Watle *(Acacia decwrens),* Dadup *(Erythrina suberosd)* and Karuvelam *(Acacia arabica)* are used as raw material in the pulp industries. To study the pulp and paper characteristics of screened chips, at a density of 224, seventy six kg/in^3 should be collected from K.MW HH 300 Chipper in case of wood and Simmer chipper in case of bamboo. They should be pulped individually in the laboratory series digester heated with indirect steam. The digester may be initially heated to a temperature of 80°C, upon which time oven-dried chips equivalent to 2000g would be charged with chemicals. The cooking should be done to get two values of kappa niuiiDer. The H-factor should be maintained at 1300. The pulping can be regulated by the H-factor, which represents time and temperature within a single variable. This concept can be applied to achieve similar pulp qualities inspite of variations due to unforeseen reasons in temperatures (or) duration of cooking. The cooking conditions should be maintained at 19-12% active alkali (as Na,O) 25%, sulphidity, 1:3 liquor to chip ratio and 90 minutes to raise the temperature to 170°C.

After cooking the chips should be discharged, thoroughly washed, defiberized and then screened. The screened pulps would be pressed to remove excess water, shredded, weighed and then sampled to determine moisture content. The pulp yield and screened pulp kappa number should be determined, and physical strength properties should be studied in the unbleached pulp. The bleaching experiments should be carried out for the pulps using CEH sequence (successive stages of chlorination, alkaline extraction, hypo chlorination). The bleached pulps then refined in a valley beater to different freeness values and hand sheets should be made to determine the strength properties after conditioning the sheet at 23 ± 1 °C and 50 ± 1 % relative humidity. All the sheets should be tested according to scan test methods. The black liquor obtained in each case should be analyzed for total solids, inorganic organic ratio, acid soluble in- organics, the presence of sodium salts (Na,O), free alkali, swelling volume ratio (SVR), calorific value and viscosity at different concentrations (Kueppen and Sitzman 1954; Palmer and Gibbs, 1967; Rys, 1949 and Singh and Sharma 1976).

Conclusions derived

Based on the above methodology our study revealed that,

a. While, mixed cook is made, a minimum of 19% alkali is required to get a pulp, having a kappa number of27±2.

b. A minimum yield of 44% can be obtained. Based on the yield percentage, if stationary digester is charged with 20 t BD chips and alkali of 18-19% keeping other parameters constant, the output would be 8.7 - 9.0 t of pulp.

c. If Karuvelam and wattle are used in excess, the same may bring down the burst factor in Kraft papers.

d. The pulp requires only 10% chlorine for bleaching to develop a brightness of 78%.

e. The black liquor of bamboo and dadup will help in the mobility of concentrated black liquor.

f. The swelling volume ratio (SVR) of dadup and wattle are good, having more than 40 ml/g. It shows that the fluidity and burning characteristics of black liquor are good if dadup and wattle are used more.

g. In all the cases, the chlorine consumption for bleaching is 9-10%. The shrinkage is more in the case of bamboo at

14%. In the case of dadup, the shrinkage is only 4%. The rest is ranging from 6-7%.

h. The mobility of the black liquor of bamboo and dadup is good since the viscosity is low when compared to others.

i. The swelling volume ratio (SVR) of all the materials is more than 25 ml/gm.

5

Estimation of the Finding of Agro Forestry Research

General Introduction

Agro forestry field experiments are considerably more complex than research confined to annual crops. It follows that evaluation in agro forestry is also a complex task. Agro forestry systems yield many products and services, all with different values and functions, and all harvested at different times over periods as long as several years. Moreover, some of the benefits of agro forestry are intangible and difficult to value, such as shade trees or live fences around homesteads.

The methods of evaluation must always be based on the objectives of the experiment. For example, the methods used to evaluate multipurpose-tree species grown on contour bunds for fodder production will be quite different from the methods used to compare different improved-fallow systems, or even from the methods used to compare the same tree species grown on bunds but for erosion control.

Researcher-managed trials are normally evaluated only in biological terms. Limited economic assessment might be possible, but research plots are usually not large enough to measure factors such as labor requirements for planting, weeding and harvesting. There can be no social Evaluation assessing the responses of farmers, if farmers are not involved in managing the experiment.

Because agro forestry experiments have so many different objectives, it is not possible to define one, or even a few, methods of evaluation. Instead we shall discuss four areas which in our experience often cause problems:

- Choosing suitable samples for measurement from experimental plots
- Converting measurements to a per-hectare basis
- Evaluating multiple outputs
- Extending evaluations over many cropping seasons.

Choosing Suitable Samples

On experimental plots that contain only crops, the border and edge effects must be considered, but otherwise it is not difficult to choose a sample area from which to measure yield.

On a typical agro forestry plot, with both trees and crops, difficulties arise from the fact that all parts of the plot are no longer equivalent. We would expect the crop adjacent to or under the trees to be different either better or worse from the crop outside the tree's influence. Some examples of measurements to assess these differences are:

- In a hedgerow-intercropping system, measuring the crop rows in the center of the alley and the row next to the hedge to determine whether tree/crop interface effects are positive or negative
- Measuring the crop in concentric circles of increasing radius to determine the distance over which scattered trees affect crop yields
- Measuring crop rows above and below terrace risers to assess the effects of trees in different position on contour terraces
- Measuring crop rows on the windward and leeward sides of trees to determine any shelter effects.

In many experiments, different parts of each plot must be measured separately. In hedgerow intercropping, for example, the evaluation of tree/crop-interface effects is based on comparing the crops adjacent to hedges with the crops in the middle of the alleys and then comparing plots with different treatments in terms of these differences. There is no reason to convert such differences to a per hectare basis.

Similarly, the pattern of crop yields at different distances from a tree can be analyzed on individual plots and then compared between plots with different treatments. Diagram E1 illustrates yield levels 1, 2, 3 and 4 meters from a tree on a single plot. To facilitate comparison with results from other plots, these data can be summarized as the difference between the two extreme values (0.38 kg/m² in this example) or as the four points (0.12kg/m²/m) or as the slope of a regression line fitted to the four points (0.12 kg/m³) (Huxley et al., 1989).

Two problems may arise. First, the researcher may fail to ensure that the crop area measured represents the system as a whole. For instance, measuring only the centre rows in a hedgerow-intercropping system will give a misleading picture of the production of the entire system.

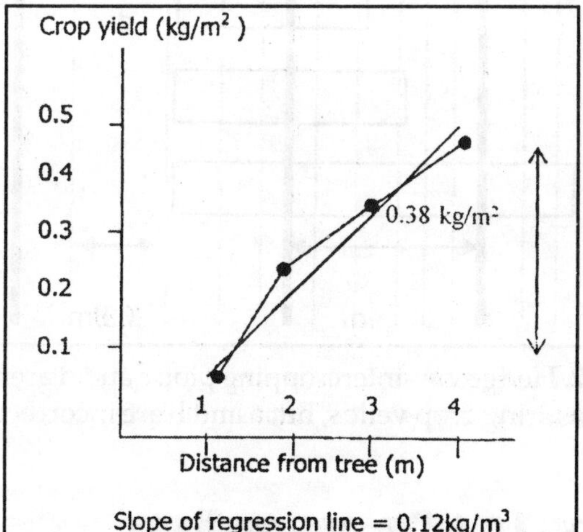

Diograme E1. Crop yeilds at different distances from a tree. The fitted regression line has a slope of 0.12/m³

Similarly, measuring crop yield under a tree and away from the tree and then averaging these two numbers will only give a correct estimate of production if the area under trees accounts for 50% of the land. If trees occupy a different proportion, then the two measurements must be weighed accordingly. The easiest way to avoid this problem is to choose a sample area that can be repeated across a field to produce a complete picture of the system.

Diagram E2 shows a hedgerow-intercropping experiment with hedges 4 meters apart and five rows of maize in each alley. Possible harvest areas are shown as a, b, c and d. Yield measurements from c or d would represent the system as a whole, although in d the harvest area for the crop is twice that for the hedge. Measurements from a or b would not give representative results as a has two middle rows and b has none.

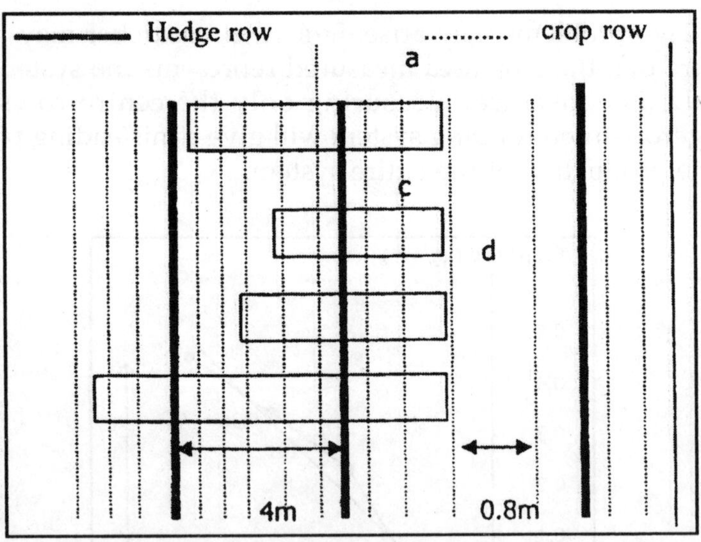

Diograme E2. Hedgerow-intercropping plot: c and d are correct sample areas for measuring crop yeilds, but a and b are incorrect.

Converting To A Per-hectare Basis

Comparisons between different agro forestry systems require calculations of yield on a per-hectare basis. For example, crop production per hectare is evaluated to compare a system of scattered trees on cropland with sole cropping or to compare hedgerow-intercropping systems using different tree species.

Researchers often make mistakes when converting yields from sample areas to yields per hectare. The problem is usually due to a miscalculation of the sample plot area, leaving out the area occupied by trees.

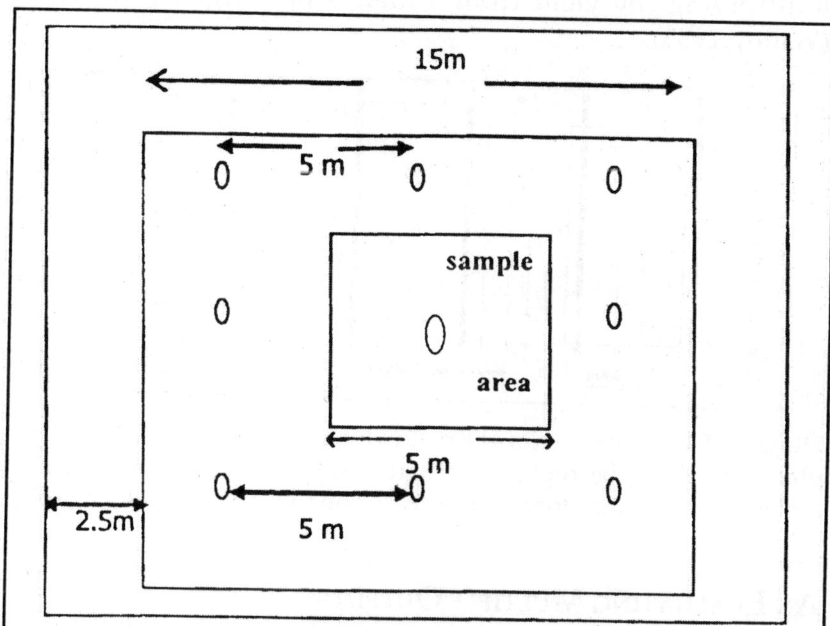

Diograme E3. Sample area for measuring crops in an agro forestry system with trees scattered on crop land at 5 meters interval. Any other sample area would not represent the crop yield the system

Again, the easiest way to get this right is to think of sample areas that can be repeated across the field. Figure 5.3 shows the correct sample areas for two hedgerow-intercropping designs plus control plots. In a, H and c, me harvest, area ii 4 x 8 meters = 32 m:. In a, only 4 rows of maize are measured. As the between row spacing is 0.8 meters, ii is tempting to take the width as 4 x 0.8 meters = 3.2 meters, and the plot area as 8 x 3.2 meters = 25.6 m². This would be incorrect because the area occupied by the hedge is ignored. Using this area would result in a per-hectare yield inflated by 25%.

Similarly, when calculating tree yields per hectare the area occupied by the crop must be taken into account. If 8 meters of hedge are harvested (Dia. E4), then the appropriate harvest area is again 32 m². Conversion to a per-hectare yield requires multiplying the yield from 8 meters of hedge by 10000/32 (Willey,1985).

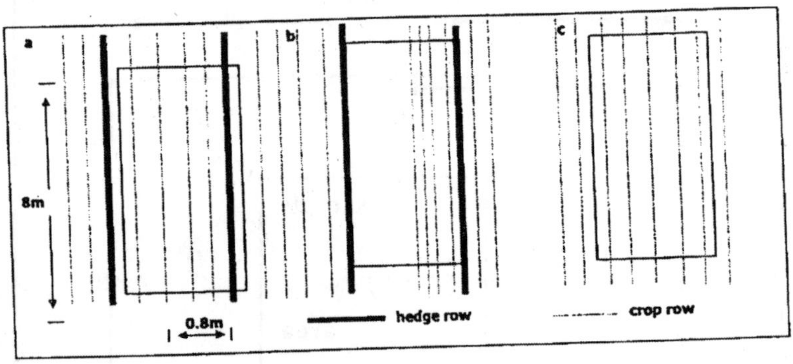

Diagram E4. Sample area of 4X8 metres for (a) a hedgerow-intercropping plot where the hedge replaces a crop row, (b) a hedgerow - intercropping plot where the hedge does not replace a crop row, and (c) a sole crop plot.

An Evaluating Multiple Outputs

In some agro forestry systems there is only one major output, such as crop yield in a hedgerow-intercropping system designed exclusively to enhance soil conservation or fertility. In such a situation, the evaluation techniques used in agronomy trials are appropriate. One technique is based on analyzing the response to fertilizer under sole cropping and under agro forestry and estimating the effect of the agro forestry system in terms of the equivalent added fertilizer Diagram E5

More common is the situation in which there are two or more useful outputs-from the tree and crop components and each must be assessed. Even if the crop component is ignored, there may be several variables related to the trees (Table E1).

No single index can integrate tree outputs (timber, fuel wood, fodder, mulch) and crop yields adequately to allow the ranking of agro forestry systems or components.

Table E1. Major Criteria for evaluating multipurpose trees for agro forestry

Criterion	Agroforestry Technology				
	Hedgerow Intercropping	**Mixed Inter-cropping**	**Fodder Lots**	**Boundary Plantings**	**Improved Fallows**
Foliar biomass (t)	x	x	x	-	x
Woody biomass (t)	x	x	-	x	x
Nutrient or protein yield (kg)	x	x	x	-	x
Depression/increase in crop yield (%)	x	x	x	x	x
Quality of foliar biomass (C.N.,N:Hgnin/ phenolics)	x	x	-	-	x
Quality of woody biomass (specific gravity)	-	x	-	x	x
Pests and disceases	x	x	x	x	x
Growth form (shape)	-	x	-	x	-
Longevity after coppicing	x	-	x	-	x

Instead, an evaluation should help explain why species differ for instance, if trees compete differently with adjacent crops because ot different rooting patterns so that an appropriate species can be selected for a particular purpose. This is illustrated by the preliminary results of an upper storey tree species screening trial in Uganda (Table E2). In this case, the species that ranked high in terms of height and diameter growth were also the most competitive with crops. Similarly, species that produced wide crowns tended to suppress crop yield.

Another complex set of agro forestry systems involves growing trees and grass in combination on contour bunds for fodder production. Different systems are compared in terms of production on the bunds including such measures as tree and grass production, total biomass and total protein produced—and also in terms of production of neighbouring crops (Huxley et al 1989).

Mead (1986) describes the value of bivariate analysis in intercropping research. This technique could also be applied to agro forestry trials that have two key outputs. It involves looking at the joint variation of the two outputs, as depicted in simplified form in Figure

E6 This can be extended to more than two variables, but interpretation becomes difficult, as graphs are hard to draw.

An obvious approach to the problem of multiple outputs is to produce an index that 'synthesizes' them into a single value. Researchers sometimes do this by assigning weights to each characteristic, reflecting an assessment of their relative importance, and averaging them. This approach can be effective if the weights are based on clear objectives, but there are pitfalls. In effect, the assignment of weights determines the results: weights can be chosen to prove almost anything!

Even if weights are based on a well-defined criterion, producing an index still reduces the information presented and can give a misleadingly simple picture. However, two approaches to indexing have proven valuable – the land equivalent ratio (LER) and economic indices.

Consider a hedgerow-intercropping system producing an annual crop plus fodder from the hedges. If both of these products are equally valuable, the system can be compared with a system in which land is divided between sole blocks of annual crops and of fodder trees in a suitable proportion. The land equivalent ratio is a convenient method to compare the biological productivity of these two arrangements by combining the two types of yield. It is defined as the land area in the sole system that would be required to produce the same yields as one hectare of hedgerow intercropping. It is calculated as follows:

$$LER = Ci / Cs + Ti / Ts.$$

Here Ci and Ti are crop and tree yields under hedgerow intercropping and Cs and Ts are yields in the sole system (Table E5).

The calculation of the land equivalent ratio should be based on the most important products of the systems. These might be grain yield of annual crop plus fodder yield of trees, or the total biomass of crop (grain + stover) and of trees (fodder + wood). By contrast, if fodder is the main product, it might be appropriate to base land equivalent ratios on protein yield.

Various methods have been proposed for calculating land equivalent ratios for even more complex systems, for example, when there are several crops per year (Balasubramaniam and Sekayange, 1990) or when the agro forestry system does not produce tree and crop components in the desired proportions (Willey, 1985). However, in such situations the land equivalent ratio has limited practical value.

The real usefulness of land equivalent ratios lies in situations where sole stands of tree and crop components are viable alternatives. If the fanner would not grow a block of trees for fodder, for whatever reason, then a land equivalent ratio that compares an agro forestry system with this alternative is not particularly useful.

Land equivalent ratios may also be calculated to assess the fodder yields of different combinations of grass and shrubs on bunds. However, it is not easy to include an assessment of the adjacent crop in this analysis since a sole-crop system—with the crop on the bund—is not an interesting alternative.

It is worth noting that the land equivalent ratio is only a relative measure: it does not reveal the magnitude of fields. Also, two land equivalent ratios are only comparable if they are both based on the same sole crop yields.

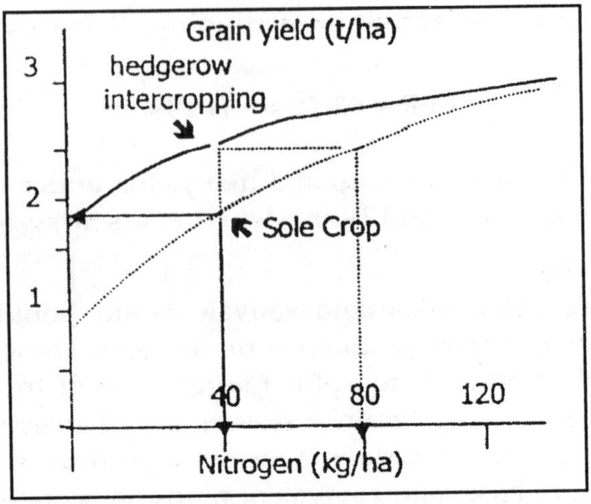

Diagram E5. Comparidon of crop response to nitrogen fertilization between sole crops and under hedgerow intercropping.

For instance, a hedgerow-intercropping system that combines *Leucaena leucocephala* with grass might have a land equivalent ratio of 1:1 compared with a system of Leucaena and grass planted separately, whereas a system combining *Gliricidia septum* with grass might have a land equivalent ratio of 1:3 compared with a 'system of separate Gliricidia and grass. However, we can not conclude that the Gliricidia / grass mixture is more productive than the Leucaena / grass system. The Gliricidia mixture may have a higher land equivalent ratio not because it is more productive than the Leucaena mixture, but because the separate Gliricidia and grass system, with which it is compared, is less productive than the separate Leucaena and grass system.

An Economic Evaluation Method

At first sight, evaluation in economic terms appears straightforward. We simply calculate each output per hectare, weigh it by its value. add up the values of all the outputs, and obtain a total value for the system. However, there are difficulties.

Table E2. Growth characteristics of seven upper storey tree species 13 months after planting and effects on adjacent bean crop.

Tree Species	Height (m)	Root Collar Diameter (cm)	Average Row yield of Bean Relative to Sole Crop (%)
Almis acuminata	1.24(7)	3.2 (6)	146(1)
Casuarina equisetifolla	2.40 (3)	3.8 (5)	100 (4)
Cordia abyssinica	3.62 (2)	9.3 (2)	46 (6)
Cupressiis lusitanica	1.61 (6)	3.2 (6)	114(2)
Maesopsis eminii	2.09 (5)	5.3 (3)	96(5)
Markhamia lutea	2.26 (4)	3.9 (4)	107 (3)
Melia azedarach	4.36 (1)	11.1 (1)	46 (6)

It is simplest to base calculations of economic value on price, but prices may vary both seasonally and in the long term. Even worse, the price of an output may be difficult to determine or may not even be relevant if there is no cash market for the product. Further complications arise from the different time periods required to produce different outputs-it takes longer to produce a building pole than a cereal crop. The costs of inputs also vary.

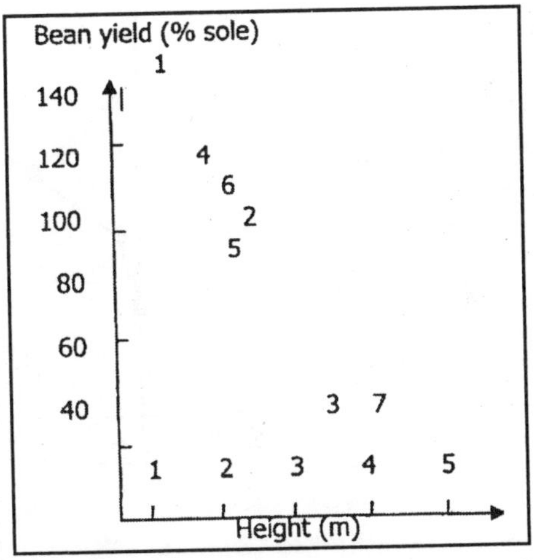

Diagrame E6. Tree heights 13 months after planting and yield of adjacent bean crop, as percent of sole crop, for the seven multipurpose-tree species listed in Table 5.2.

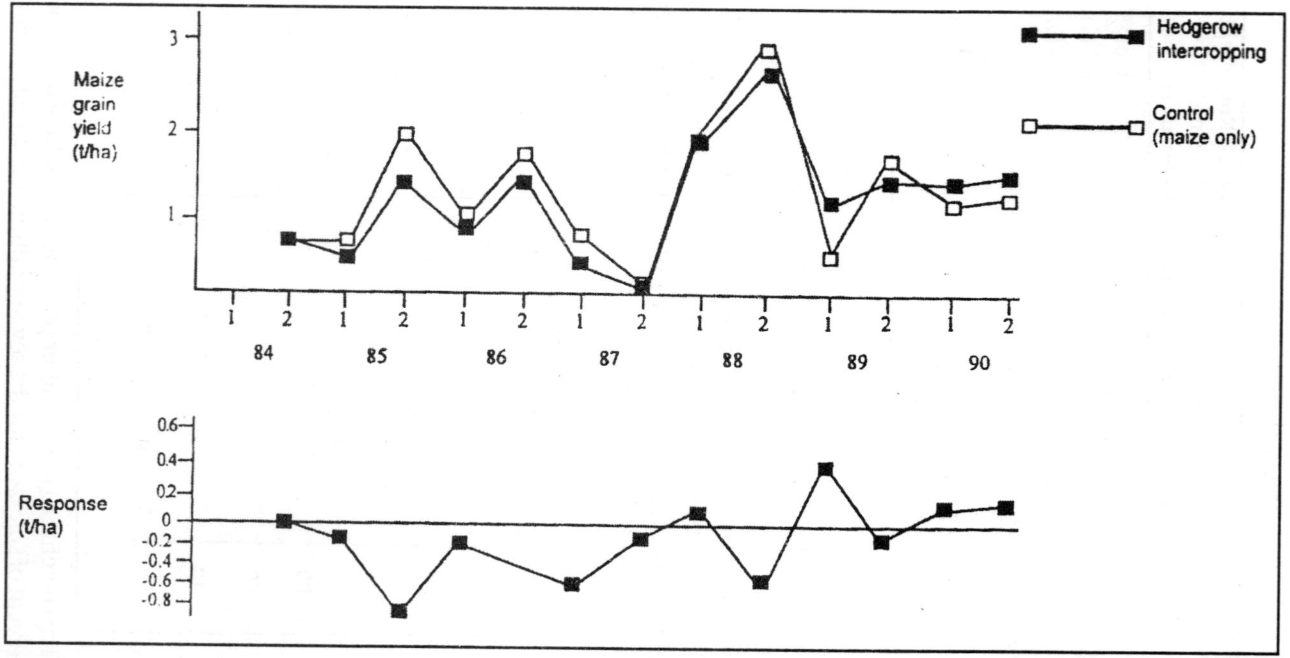

Diagrame E7. Upper graph shows average maize yields at Machakos, Kenya, under hedgerow intercropping and sole cropping during 14 growing seasons, from 1994 to 2000. Lower graph shows the response of maize yield to hedgerow intercropping. Values below zero indicate that crop yields were lower under hedgerow intercropping than sole cropping, whereas values above zero indicate that yields were higher under hedgerow intercropping (Mugendi and Muthoka, 2001).

Evaluating Over Many Seasons

It is essential to evaluate agro forestry systems over a sufficiently long period, typically for a minimum of five years. There are two main reasons for this. First, many of the potential benefits of agro forestry are long term. Obvious examples are soil-erosion control and fertility maintenance. Whilst it might be possible to measure some of these benefits directly over a few seasons, their impact on crop productivity must be monitored for a longer period. Trends can be depicted simply by plotting outputs from different systems over time (upper graph in

The second reason is the variation in weather that inevitably occurs from year to year. The relative performance of different agro forestry systems can be quite different under different weather conditions. A system that performs well when there is adequate rainfall might perform poorly in seasons of water shortage.

It is difficult to give a rule for the number of growing seasons necessary to evaluate different systems, but the evaluation period must include the range of climatic conditions usual for the site. The graphs shown in Diagrame E7, will change according to variations in the weather, but trends will become apparent over time.

Some seasonal variation can be removed by plotting responses-the differences from a standard control—as shown by the lower graph in Diagrame E7. However, such calculations do not give absolute values, which are important when considering sustainability. Plotting responses may also obscure a pattern in which one system does well in good years and another does we!! in poor years.

When a particular climatic event has an important effect on trees or crops, we can calculate the likelihood of it happening again using long-term meteorological records. Suppose tree establishment is poor because of a three-week dry spell after transplanting, the chance that such a dry spell will occur in any year can be calculated simply by looking through climatic records. If such a spell occurred only twice in the past 40 years, we can estimate the risk as 2/40 = 5%, probably an acceptable rate. If the risk is 75%, then any agro forestry system that would fail in such a dry spell will not be viable.

Table E3. Land equivalent ratio of Leucacena hedges producing fodder (dry weight) with a maize

Sole Maize	Sole Leucacena	Hedgerow Intercropping
4 t/ha grain fodder	6 t/ha fodder	3.2 t/ha grain + 2.4 t/ha

$$\textit{The land equivalent ratio of the system} = \frac{3.2+2.4}{4.0+6.0} = 1.2$$

This means that the productivity of 1 hectare under hedgerow intercropping is equivalent to that of 1.2 hectares under the sole system. In other words, hedgerow intercropping is 20% productive than the sole system.

However, differences in weather could still affect the growth of The fallows themselves. Thus, if other factors are under consideration in addition to length of fallow-tree species for example—then the entire experiment needs to be repeated starting in different years. If the design does not account for the effects of weather in different years, then the best that can be done is to look at responses, comparing various treatments with a control. Even so, the confounding effect of weather cannot be eliminated completely for two reasons; the magnitude of responses varies according to the weather, and the actual effects of weather may not be proportional to the way weather is measured normally based on the amount of rainfall.

During fallow periods, land is out of production. Particularly in areas where land is scarce, it is important to compare different fallow systems on the basis of average annual production taking both the fallow and cropping periods into account.

The situation is further complicated if the tree fallows yield some economic products such as fodder, fuel wood or timber. These products must also be considered in an overall assessment of the system.

Under diverse and rapidly changing conditions—particularly socio-economic conditions—no one evaluation method can give all the answers. An agro forestry system that works well in one location may not be appropriate in another, and a system that works well today may not be appropriate tomorrow. In addition, if systems are judged purely according to short-term results, important sustainability aspects may be overlooked (Mugendi and Muthoka, Rao and Roger,

A comprehensive evaluation must consider the economic and social aspects of an agro forestry system, as well as the biological. Such an evaluation requires accurate long-term data. Experiments should be conducted over a minimum of five years even longer if possible. A shorter experiment not only misses out the long-term potential of an agro forestry system, but also risks inaccurate conclusions due to short-term vagaries of climate.

Finally, the difficult task of integrating the effects of soil, climate and various plant-management options calls for the development of appropriate software development and computerized models. Models could help assess resource capture and utilization, nutrient cycling, the growth and productivity of components, and long-term sustainability. This is an area where considerable work is yet to be done.

6

Tcchniquc Uscd in Social Forestry Research

General Introduction

A Social forestry programs in India have grown in importance and scale and now continue to be a major element in India's overall program of rural development. From modest beginnings over a decade ago, there has been an almost exponential growth in the human and financial resources devoted in social forestry.

The term social forestry is difficult to define precisely, but is generally understood to mean tree-growing (including associated products, e.g., bamboo, grasses, legumes) for the purpose of rural development. As social forestry has a rural development focus and is heavily dependent on the active participation of people, it is also known as "forestry for local community development" or "participatory forestry" (FAO, 1998, 2000). Although wide ranges of activities are included in social forestry, five main components can be distinguished in India. With variation, they

A. Farm forestry (tree growing on private land),
B. Farmer leasehold or tree patta,
C. Village woodlots or community forestry,
D. Strip plantations alongside roads, canals, railways, etc., and

E. The reforestation or rehabilitation of degraded forest areas.

The Social forestry programs mainly used for one or more of these components. There are also distinctions between and within these components depending on who have owns the land on which the trees are being planted (e.g.,. private farmers, private industries, municipalities, forest departments, revenue departments, etc.) or who is responsible for the planting (e.g. farmers, villages, cooperative, voluntary agencies, rural development departments, Panchayat schools, etc.).

Although these distinctions are sometimes blurred, all forms of social forestry are differentiated from commercial or territorial forestry by the involvement of the rural population in decision-making, management and as beneficiaries.

The farm forestry is tree planting undertaken by individual households on their own land or land they have rented from others. Tree seedlings may be planted in blocks (small plantations), on field boundaries, or around homesteads. They may be intermixed with agricultural crops in several forms of agroforestry, or they may be planted alone on either agricultural land or uncultivable wastelands. Farmer leasehold or tree patta denotes a kind of farm forestry in which poor farmers or landless labourers are given leases to tracts of public land on which, with varying degrees of public support, they are constrained to grow trees. Village woodlots are small plantations on communal or government land, operated by or on behalf of the village, for the benefit of the village as a whole, although there may be special arrangements, which provide preferential treatment to the under-privileged. Strip plantations are relatively narrow areas along the sides of roads, canals, railways, and rivers, established by the Government (usually the Forestry Department) with the intention of providing the benefits of forest products to local people and to serve as demonstration areas. The reforestation or rehabilitation of degraded forest refers to large plantations on public lands, which have been degraded, and which are often in environmentally critical areas. Such plantations may or may not be considered a form of social forestry depending on whether or not there is significant involvement of local communities (1998)

The objectives of social forestry necessarily differ by components. While all social forestry aims to increase tree production and reduce environmental degradation, the nature of the product, the type of management, and the distribution of benefits depend on the type of social forestry involved. Farm forestry is designed to help rural households better meet their own needs, whether through the direct production of fuel wood, fodder, and poles for their own use or through the production of a commercially marketable crop of poles or pulpwood. Tree patta forestry is similarly designed to increase the incomes of poor households through the sale of forest products and at the same time to help satisfy their need for fuel wood and fodder (1999a) Village woodlots are intended to provide tree products, particularly fuel wood and fodder, for the community as well as (in many cases) incomes to the local village panchayat. Strip plantations and reforestation are designed to provide local communities with some of their fuel wood and fodder needs and to conserve and improve the environment. To a varying degree, each of these components has features, which deliberately target benefits towards the poorest and most under-privileged sections of society, including rural women who are frequently those hardest hit by the growing scarcity to tree products. However, these poverty-alleviation objectives and those related to production are often confused and this contributes to the widespread controversy surrounding the social, economic and environmental effects of social forestry (1999).

The massive social forestry programs being carried out in India are, at present, spearheaded by the National Wastelands Development Board specially created by the Prime Minister. While the principle implementing agencies for these programs in each state are the Forest Departments other government departments and private and voluntary agencies, are increasingly being mobilized to meet national objectives. State Forest Departments have typically introduced new organizational structures to pain and implement social forestry activities and have appointed separate staff in the field and at headquarters. For forestry departments as much as for other departments and agencies, social forestry representsa departure from traditional activities and styles of working (1999).

TGUIDELINES FOR SOCIAL FORESTRY PROGRAM

A Community forestry programs are based upon growing trees on public or community land as opposed to private farms. The degree of local participation in planting and looking after the trees varies. What all community programs have in common, is that they are intended to provide benefits, which are shared by the community as a whole.

A most common type of community-forestry programs is that in which the forest department takes on the responsibility for carrying out the planting. The community provides inputs such as fertilizers and seedlings without any outlay. The engagement of the local community in the implementation of the schemes of this type is largely passive and is normally restricted to the provision of hired labor for planting and an agreement to cooperate in protecting the plantation.

On Other programs rely on a much higher level of community participation and control. They are generally designed to use land, which is under direct community ownership; (or) they can take place on state lands, which have been specially designated for community control. Community itself takes the main responsibility for planting and looking after the trees and the role of the promoting agency is primarily a catalytic one. Community programs can use both commercial and non-commercial incentives as a way of stimulating local participation. In all cases, the key to success lies in persuading local people that the program is in their interests such that the benefits they are promised are securely guaranteed.

Degree of local people participation

It was discussed earlier, the aim of community forestry programs are to regenerate the degraded forests and barren lands.

(i) Local institutions involvements

A Community tree growing programs are crucially dependent upon the active collaboration of village councils; community groups (or) other local institutions. Program planning must therefore be based upon a clear appreciation of actual role, such organizations play in local affairs, and their potential for realizing or thwarting the aims of the program. The village community groups should be trained or giving proper instructions, about the growing of bamboo in the barren land by the specialists in the field. While selecting the village community groups, preference will be given only to the educated unemployed youth.

(ii) Land required in Programme

Maximum 30 ha area of degrade/barran forestland is to be required at a time to each participating village group. In future, more land can be allocated if the work of the participating village committee is found encouraging.

(iii) Adopted Procedure of working

After collaborative micro planning a need-based management plan will be drawn up. The forest department will assist the village committee to establish joint protection and management systems

(iv) Requirements/Rule of working

Order requires participating village committees to project the allocated land against non-forestry use, encroachments, grazing, illicit-felling and wildlife. Land in no case is to be allocated to individuals and ownership of the land shall remain with the Government. It is also required to execute an agreement bond-between forest officials and the participating village committee.

(v) Resource sharing

Entire bamboos obtained from community forestry programs will go to the members of the participating village committee. At the time of harvest prescribed in the management plan, after deducting the government expenditure, 60 percent share of net income will go to the members of the participating village committee and 40 percent share to the state exchequer. Participating village committee will again invest at least 50 per cent of this income in plantation activities.

(vi) Monitoring of works regularly

A supervisory committee is to be constituted to monitor the development, protection and management works of the community forestry. Members of the committee will include respective district forest officer, a range officer, and a representative of village committee members.

Limitations of lands

(a) The problem of land- use conflicts

After obtaining the necessary land for tree growing is undoubtedly one of the major stumbling blocks in programs of this type. Community land is frequently scarce, or is being used for a variety of other purposes. Agreement to dedicate it to tree growing can be very difficult to achieve.

(b) Lack of identity of interests

A few communities have a genuine identity of interests among all their members. Practical experience has shown that this can cause a number of severe problems with community programs, where communities are rigidly stratified along social, economic or religious lines, the barriers to communal action can be partially difficult to over come.

A Cope for Community Forestry

The Community forestry programs, with their needs for compromise and reconciliation of opposing interests, means that they will always require a great deal of careful preparation and negotiation. This demands a high level of trust and co-operation between the community, and the promoting agency, which can, itself take a long time to create.

Monitoring and Evaluation of Programme

In despite of the magnitude of the social forestry programs in India, there is little reliable information available to policy makers and program managers regarding the progress and effects of the program (NWDB, 1989b). This is perhaps, because of the novelty of the approach and the rapid and recent expansion. The widespread lack of effective information systems has hampered management and policy makers in improving the design and implementation of their programs. The scarcity of reliable information has also led to controversies and debates fueled more by impressions, isolated field visits and pre-determined opinions than by objective empirical information.

On reasons given, the Government of India and most of the State Governments and other agencies involved in promoting social forestry have placed a high priority on developing an effective monitoring and evaluations (M&E) system to serve both the State Government Departments and Agencies and the Central Government. To meet this need, most social forestry projects have made specific provision for the establishment of monitoring and evaluation units (MEUs), which are expected to collect and interpret information on all social forestry activities. In a number of states, such units have been set up and have commenced operations. In other states units are planned

The primary role of monitoring and evaluation is to improve program or project implementation so that the program's ultimate objectives can be met and/or periodically measure progress towards these objectives

Monitoring can be regarded as an assessment of the efficiency with which the program is implemented-including measurement of the quantity and timing of input delivery and output produced. Monitoring is usually understood to include the tracking of both financial and physical activity through regular quantified reports. While detailed financial reporting and accounting systems are usually already in vogue for auditing purposes, systematic physical reporting and monitoring of the implementation process (both problems and progress) is not well developed. The relationships between financial expenditures and physical output are not usually studied and hence little is known about the unit costs of achievements or whether lower costs would have produced equivalent results (2001).

In contrast, evaluation can be construed as the assessment of the results of implementing the program. On going evaluation is a means of examining the most important direct effects of the program concurrently with this implementation. This form of evaluation, sometimes called beneficiary monitoring, is meant to serve as an extension of the monitoring system in the sense that its primary purpose is still to provide managers and policy makers with information about the individuals and communities affected which will allow program implementation to be improved

Ex-post evaluation is differentiated from on going evaluation being conducted when a project or program is completed as a means of assessing its overall impact and achievements, This type of evaluation, while not inherently undesirable, is complex, has a rude appetite for data, and usually requires sophisticated analytical methodologies and considerable financial and computing resources

Since monitoring and on-going evaluation are closely related they can often profit from being functionally integrated. Hence, while monitoring lays stress on the reporting of project progress and constraints as a time bound aid to program management, it should be closely linked to evaluation when the later is used as a means of determining whether implementation is meeting its original objectives or producing unanticipated results which may affect (positively or negatively) the outcome of the program.

A widely recognized in India, a separate and specialized monitoring and evaluation unit (MEU) is necessary to implement an effective M&E system. In order to maximize the MEU's direct usefulness to management and to the daily implementation of a social forestry program, it is recommended that, as is generally the practice, the unit be located within the implementing department or agency and report directly to top management. The duties and responsibilities of the MEU can be summarized as follows.

(i) To establish, in cooperation with program management, clearly defined objectives and regrets for program implementation against which progress can be monitored;

(ii) To implement and operate a monitoring system, including the development and application of methodologies and procedures for the collection and analysis of information.

(iii) To collect information to enable the periodic evaluation of progress and effects through existing administrative and accounting records, surveys and studies and to analyze, interpret and report the findings to management and, through them, to policy makers and other interested bodies;

(iv) To undertake on ad hoc basis, enquiries and studies in order to solve urgent problems for management; and

(v) To follow-up on the recommendations and results of monitoring and evaluation and assist in integrating them into future plans.

However, MEUs should not generally undertake investigations that are more properly classified as research studies, even though they may be relevant to program performance. For example, studies of the impact of different silvicultural regimes on water tables, of the agronomic and economic interactions of trees and crops, or of the impact of increasing fuel wood supplies on health and nutrition should all be considered as long term research studies outside the purview of an MEU. Not only are such studies long-term in that they have to be carried out over many years for the results to have some validity, but they are costly and technically complex and are, therefore, best carried out by research institutions.

An Overview of the M&E System

By aiding the understanding of the system as a whole, Diagrame F1, provides an overview of the sources, flows, and destinations of M&E information. The top most part of this figure shows the main activities of a social forestry program. The categories of plantations are shown in Diagram F.1 which deals with progress reporting. Downward opening triangles denote monitoring information, derived from nursery registers and plantation journals, which is reported for all activities by social forestry field staff. Upward pointing triangles indicate data collected by the MEU through sample surveys and special studies which comprise a different, but important part of monitoring and evaluation. This part of the data collection system is primarily required for "People's Plantation" where departmental staffs are not directly responsible for planting activities and thus cannot be expected to report on them as part of their normal work

The lower part of Diagram F.1 shows how the data obtained by the MEU after being transformed into usable information and summarized in reports, the most concrete output from monitoring and evaluation units, is disseminated to managers, policy-makers, and the field personnel responsible for implementation in order to provide feedback to improve performance.

For obtaining truly useful result monitoring and evaluation must be undertaken in a thoroughly professional manner by specialists provided with adequate resources. This is only possible if the M & E unit is integrated with, used and directed by program management. Hence, it cannot be over-emphasized that management must fully understand the purpose of monitoring and evaluation and resist any temptation to regard M&E staff as an inspectorate or internal police force whose

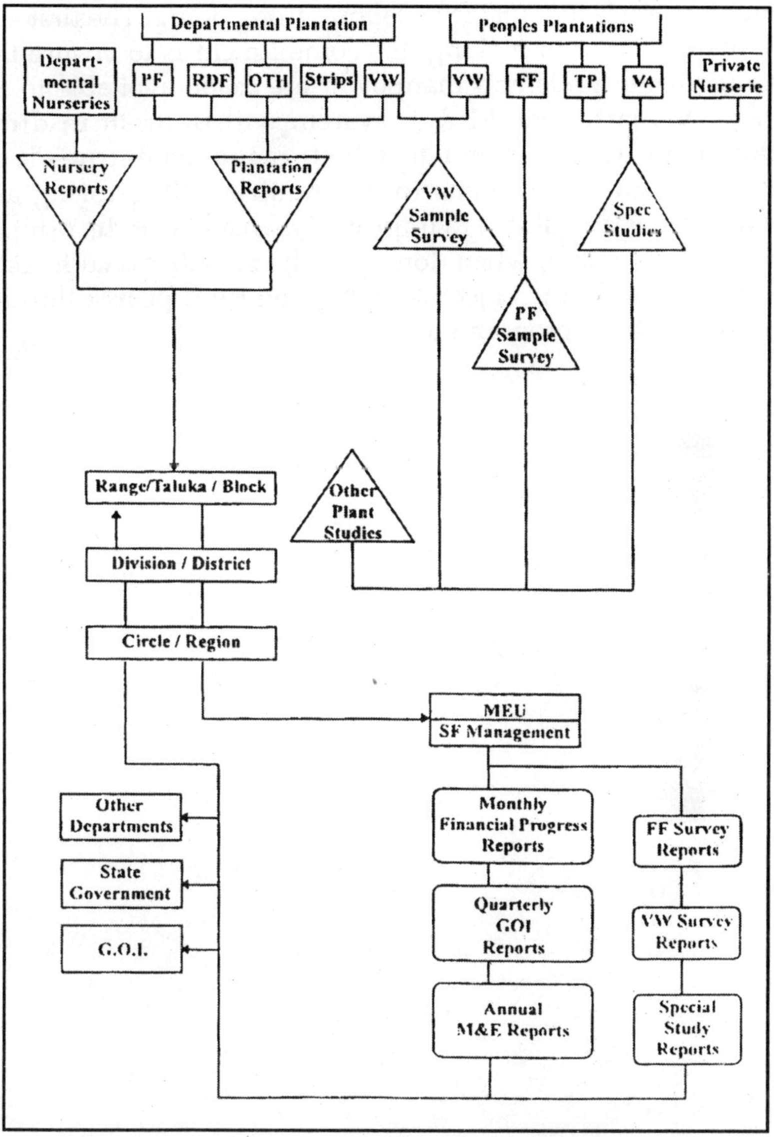

Diagrame F1. An Information Flow in the M&E System

Summary: Production Forestry = PF, Rehabilitation of degarded Forests = RDF, Other = Oth, Village Woodlots = VW, Farm Forestry = FF, Tree Patta = TP, Voluntary Agencies = VA, Special = Spec, Social Forestry = SF, Monitoring and Evaluation Unit = MEU.

sole purpose is to criticize or report wrong doing. The first step by management in making this commitment is to ensure that the requisite funds and manpower are made available to the unit. Thereafter the M & E system will flourish in direct proportion to the extent to whicli management needs its products and gives focus to its enquiries. Monitoring and evaluation is an adjunct to improved management. Its primary justification is that, when done well, it can help social forestry programs attain the objective of helping rural people through tree technology programme.

7

Technique Used in Nursery Development

General Introduction

There is no any little doubt if any about the need to promote afferstation in the developing world. Governments and peoples of developing tropical and subtropical nations worldwide are recognizing the ecological and economic value of maintaining and restoring their forested lands. Vast efforts and sums are being expended by governments, organizations and individuals throughout the tropics to protect existing forests and to reforest, afforest, or otherwise rehabilitate lands denuded of trees. Hundreds of millions of tree seedlings are being produced and planted annually across the tropics, resulting in the establishment of many square kilometers of new forests, or the integration of trees into the farming systems of millions of small land holders.

Un fore seen the impacts of these enormous expenditures of human energy and energy financial resources are being severely hampered by the use of antiquated, and in some cases, deleterious, methodologies and technologies. Many of the tree seedling production systems currently being used in the tropics are based on outdated technology, and these are the focus of this chapter. Nurseries are the cornerstones of nearly all reforestation success. Yet nurseries in the developing world nearly always depend upon labour-intensive seedling production methods that require long time periods in the nursery which increase costs, hamper seedling transport and thus restrict large-scale tree seedling distribution and planting.

The tremendous opportunities exist for the improvement of tree seedling production systems to provide greater numbers of more vigorous seedlings at significantly lower costs. There are two separate aspects to seedling quality the genetic constitution and the physical quality of planting stock. The following sections deal almost exclusively with the physical quality of containerized seedlings. Plants with improved survival and growth create better financial and ecological returns, and can generate sustainable enthusiasm and support for institutions to continue or even expand their reforestation programs. In other words, success breeds success, especially with reforestation programs.

It endeavours to draw together the combined experience of nursery managers and specialists worldwide, specifically focusing on the adaptation of currently available root trainer production systems to the unique conditions existing in tropical developing countries.

Systems of Nursery Production

Nursey production In the tropics: Improvement of roof

The tropical and sub-tropical climates are ideal for the establishment and growth of forest tree species. Reforestation with agro forestry program, the decision-makers are faced with a wide array of potential sources of forest tree planting stock for use in their projects, including bare root seedling production, aerial or direct seeding, and containerized nurseries.

In contrast to nursery production in the temperate zone, bare root seedlings are not generally produced in large quantities in the tropics because of the greater heat, moisture, and transpiration stresses.

The raising for high quality nursery stock with root trainer systems in the tropics

An intensity of management of root trainer nurseries is quite different from that required when using poly-bags. Seed sowing methods and dates, irrigation, fertilization, and other cultural regimes must be changed significantly if quality results are to be obtained. The entire nursery staff will have to be retrained in the more precise and demanding aspects of root trainer nursery management. Experience shows that it is clearly possible to do so and there are many examples of successful root trainer nursery production in developing countries worldwide.

A production system through Choosing a root trainer tree seedling

An improved root growth and morphology are important goals of any carefully designed root trainer system. Nursery managers and reforestation program directors can feel comfortable that most root trainer systems currently on the market can produce a vigorous seedling with good root growth and orientation, given a good growing medium, proper care and management in the nursery.

To maintain the proper size root trainer project managers must first determine what size seedling is desired because the size of the root trainer influences the size of the seedlings produced. In general, the bigger the container, the larger the potential size of the seedling, and the longer the time required in the nursery. When deciding on seedling size the degree of competition from weeds and other plants upon out planting, the soil depth and degree of drought-hazard at the planting site, and the particular tree species spatial requirements should be taken into account. For example. Wood (1960) reports that the ideal container size for producing eucalyptus is approximately 90 mm deep, 44 mm wide at the top, with a volume of about 90 cc, and a seedling density of about 420 plants/m^2.

The good weed control on the planting site generally allows for the use of smaller seedlings (and subsequently nursery reduced costs). If the site to be planted has poor weed control, larger seedlings (and thus larger root trainers) are usually required. However, there are occasional exceptions..For instance, relatively small and cheap seedlings produced in root trainers with volumes of approximately 89cc averaged an acceptable 60% survival after one year in farmer-managed agro forestry plantings; even with poor weed control (Whitcomb, 1983). Larger (and more expensive) seedlings may not have fared any better since the majority of the mortality was due to damage from cultivation of immediately adjacent crops by hired labour

The Density of seedlings in the nursery is important. Broadleaf seedlings which are crowded together in root trainers will grow more slowly become spindly and may suffer from increased disease due to high humidity and poor air circulation (Stoecke ler and Jones, 1957). Stem diameters will be smaller, which may reduce out planting survival and growth (Rao et al, 1990). While Casuarinas and Pine seedlings can be closely spaced with little problems other species, such as coffee, cedar, mahogany and teak, need much wider spacing. Block type root trainer systems only allow changes in spacing when alternate cavities are unfilled; this would be needed to allow seedlings grow larger during the season. Containers with removable cells or tubes, or book_type containers permit mid-season corrections in spacing according to the particular cultural requirements of the species being produced.

Lastly, if soils on the planting site are shallow, or if the site experiences occasional drought even during the planting season, then choosing a root trainer with deeper cells can be advantageous. Longer plugs can be planted deeper in the soil to take advantage of soil moisture in lower zones, and thus allow the seedlings to better survive in periods of drought (Peterson and Smith 1975).

Summarizing managers deciding on a root trainer system should focus on issues such as the desired size of the seedling at out planting, flexibility of the system to adapt to different cultural requirements of a number of species, the availability and cost of labor, the ability of the root trainer system to use manual procedures, the short and long-term costs, and the current availability or potential for manufacture within the particular country.

Control of Moisture

The Control of moisture is critical to proper seedling development, and also influences insect and especially fungal pathogen populations in root trainer nurseries. Root trainer systems that use well-drained and well-aerated polling media often need to be watered more often than poly-bags. But because of the close spacing and small cell volumes, root trainer nurseries use one third to one half the amount of water required in poly-bag nurseries of equivalent production (Perkins, 1973). Over or under watering can be serious problems when nurseries shift from using poly-bags to root trainer systems. Since the small plugs dry out quickly, nursery managers must be trained to check their seedlings twice daily and to irrigate only when needed, not on a schedule. This helps to avoid over or under-watering. Root trainer systems that can be opened to enable easy inspection of soil moisture are preferred.

While shade cloth-covered shade houses have better air circulation and lower air temperatures than those using plastic films, plastic has been found to be superior for reducing disease incidence, especially when the nursery season coincides with the rainy season. Plastic permits the nursery person to completely control the moisture content of media in the root trainers, even during periods of prolonged precipitation, greatly reducing the incidence of soil-borne pathogen attacks. Shade cloth cannot exclude precipitation and forces the manager to heavily rely on fungicides to control soil-borne pathogens. Bamboo and palm leaves are also frequently used for shade in the tropics because they are readily available and cheap but they should be avoided whenever possible, however, as they do not exclude precipitation, and also because they harbor insects and pathogenic organisms.

Cultural Approaches

Since root trainer containers are much smaller in volume than poly-bags and intentionally include features designed to maximize growth, it takes a much shorter period of time to produce a seedling, usually only 3-4 months for most species, half the usual length of time. Thus, careful timing of seed sowing is critical if root bound and stunted seedlings are to be avoided.

Seedlings in small containers must be fertilized more frequently than those in larger containers (Whitcomb, 1983). Some root trainer nurseries normally utilize a balanced liquid fertilizer drench of 20-20-20 NPK several times a week for the first 6 weeks or so of the season, followed by a 9-45-15 drench later in the season to encourage root growth and slow height growth (Perkins 1973). Manipulating soluble fertilizer formulations, concentrations, and application frequencies allows the nursery manager to better time the maturation of the seedling with the onset of the rainy season, a distinct advantage. Micronutrient drenches are applied, as needed, usually once every 2 or 3 weeks. Other nurseries mix slow-release fertilizers into their potting media. Nursery manager often must adopt their fertilization regimes to the pH of the irrigation water, rather than adjusting the water pH, since most nurseries in the tropics lack the equipment and infrastructure necessary to change the irrigation water pH. Specific nutritional requirements for many tropical species have not yet been determined, resulting in over fertilization and excessive height growth in some species and under fertilization and nutrient deficiencies in others.

Proper transplanting of newly germinated seedlings into root trainers is essential if the root training features of the container systems are to be utilized. Poor transplanting (Pricking-out) techniques can result in very serious root deformations causing poor growth and stability upon out planting.

Newly emerging seedlings should be pricked out at the cotyledon stage before the formation of many roots, thus avoiding bent, inverted roots. Personnel transplanting seedlings should be well-trained and closely supervised.

Due to the small size of the containers, root trainer nurseries are somewhat less than other types of nursery production. Indeed, "failure to recognize the fundamental principle that container seedlings require intensive management will inevitably lead to failure and negate the promise of consistent and reliable production of high quality seedlings". Programs intending to use root trainer systems must have sufficient managerial resources to adequately operate the nurseries. Thorough and comprehensive training of all nursery personnel to assist them in making the transition from cultural approaches used in poly-bag or bare root nurseries to those used in root trainer cultural systems is absolutely essential if consistent, high quality seedling production is to be achieved.

Packing, Handling & Transport

Most root trainer nurseries achieve their goal of producing healthy, vigorous seedlings ready for out planting by a predetermined date, usually at the time the rains begin. Yet the nursery's fine work is often squandered because of serious damage inflicted on seedlings when removing from the container packing, transport, and out planting activities. During this final production phase, which ends at the planting site, formally vigorous seedlings are routinely yanked from their containers, crushed, dehydrated, overheated and suffocated in boxes, and left to suffer in the Sun for long periods prior to planting. This abuse inevitably leads to reduced survival and growth and represents large financial losses.

Measures must be taken to protect seedlings on their trip from the nursery to the planting site. Seedlings should be watered thoroughly the night before packing. If seedlings are removed from the root trainers in the nursery, it should be done gently, protecting the plough from root damage and loss of soil. Packing should always be done in the shade to avoid heat and moisture stress. While some programs (e.g. New Zealand Forestry Programme) pack the seedlings in alternating layers in a cardboard box lined with a sheet of plastic, other programs utilize wooden boxes or plastic bags. This approach can result in loss of soil from the root plug from abrasion and vibration during transport, poor aeration, and overheating of the seedlings. Seedlings shipped in this manner must be planted immediately upon arriving at the planting site, regardless of the weather conditions. A method developed in Haiti is to pack the root trainer grown root plugs in a plastic bag (25 seedlings or so to a bag, with the leaves emerging from the bag), and then put the bags in a box. Using this system, soil adheres better to the root plug and crushing and abrasion damage is minimized.

When using containers with portable cells, such as the Ray Leach Single Cell System, Hawaii Dibbling Tubes, etc., the seedlings can be transported to the field protected within their containers, a major advantage since physiological deterioration of the seedling is avoided. They will require more transport space however, and it may be difficult to collect the containers after out planting is completed, especially in agro forestry programs working with small farmers.

Finally, the very best seedlings can be rendered useless by poor planting methods, poor site preparation and weeding, and poor post-planting care and protection. Any nursery program must be part of a larger program emphasizing proper implementation of these basic reforestation techniques.

Th nursery in tropical developing countries are currently producing hundreds of millions of trees every year, mostly using antiquated methodologies and techniques. Seedling production in polybags is labor and space intensive and produces seedlings that are large, heavy and difficult to transport. They frequently have severely deformed root systems which will take time to function properly, which may compromise growth in later years, significantly reducing the production capacity and potential environment on economic impacts of any reforestation or agro forestry program.

An obvious by project and nursery managers wishing to improve the cost effectiveness, efficiency, and productivity of their seedling production operations should seriously consider switching to root trainer tree production systems. Better root growth, form, and vigor avoidance of spiral roots (with its long term investments implications) simplification of insects and disease management, lower labor costs, smaller quantities of potting media, reduced cultural, handling and transport carts, and better initial survival and subsequent growth of out planted seedlings, are persuasive arguments for change. Clearly, the modest initial investment required to change to root trainers can result in large gains in seedling quality, productivity, and cost effectiveness (World Bank, 1993).

Low-technology Techniques for the Vegetative Propagation of Trees

Stem cuttings of have been easily rooted in improved low technology, high humidity polythene propagators. These propagators, which are cheap to construct, arc very effective and have no essential requirements for either piped water or an electricity supply. Experiments have tested different rooting media, auxin applications and compared mist versus non-mist propagation.

An assessments of the physical and gaseous environment of the propagators has indicated ways of improving the rooting environment through and understanding of the sensitivity of the relative humidity to radiant energy and to opening the propagator for short periods (e.g. 2-3 minutes).

Now it is widely realized that vegetative propagation and clonal selection offer a means to greatly enhance the yield and quality of forest products from commercial plantings in the tropics (Williamson, 1983). However, there is a need to simplify the technology so that vegetative propagation can be achieved in the absence of electricity and a piped water supply. In addition, in many tropical countries, the high capital & running costs of currently available propagation systems make them inappropriate except for research or large-scale commercial projects.

An environmental requirements for root initiation in leafy stem cuttings are those that minimize physiological stress in the cutting. In general terms this means using shade to lower the air temperature and, by providing a high humidity, to reduce transpiration losses. By the latter, the vapour pressure of the atmosphere surrounding the cutting is maintained close to that in the intercellular spaces of its leaf.

There are numerous propagation systems used in commercial horticulture. These are usually based either on spraying mist, fogging or enclosing in polythene. The advantages of polythene systems have been known for many years and they have been used to propagate tropical hardwoods with good success.

Recent work by the Institute of Terrestrial Ecology (ITE) and its overseas collaborators has applied and improved the design of non-mist propagators for use with a wide range of timber and multi-purposes tree species from both tropical moist forests and semi-arid areas. Recent studies with *Triplochiton scleroxylon* cuttings under intermittent mist have indicated that rooting ability is related to the production of reflux - soluble carbohydrates, apparently derived from current photosynthesis while the cuttings are in the propagation unit. Furthermore, it seems that the ability to produce these carbohydrates is related to the pre-severance light environment and nutrient status of the cuttings while on the stock plants. Both the total irradiance and the light quality (red: far-red ratio) are important components of the pre-severance light environment and these factors interact with nutrient availability to influence the rates of net photosynthesis and rooting. These variables to a large extent account for the variation in the rooting ability of cuttings of *T. scleroxylon* taken from different shoots of variously-treated stock planes

Juvenile shoots of twelve tree species (Table G.1) have been used as leafy stem cuttings. The studies presented here were done in either glasshouses or under nursery conditions. In all instances however. 22-27 °C the cuttings were prepared as described below and set in randomized block. The numbers of replicate cuttings per treatment were between 24 and 117. Standard errors for percentage rooting were calculated using the procedures of Barber (1984) for data with binomial distributions.

Table G1. Tropical tree species vegetatively propagated using simple, low-tech propagators.

Species	Family	Uses
Gmelina arborea Roxb.	Verbenaceae	Timber
Eucalyptus deglupta BL.	Myrtaceae	Timber
Nacuclea diderichii (Dc Wild & Th Dur.) Mcrr.	Rubiaceae	Timber
Lovoa trichilliodies	Meliaceae	Timber
Harms Ricinodendronheudelotii Pierre ex	Euphorbiaceae	Fruit
Cordia alliodora (Ruiz & Pav.) Oken	Ethetiaceae	Timber
Vochyasia hondurensis Spargue	Vochysiaceae	Timber
Albizia guachapele (Kunth) Dug	Mimosaceae	Timber
Prospois juliflora	Mimosaceae	Multipurpose
Acacia tortilis (Forsk.) Hayne	Mimosaceae	Multipurpose
Terminalia spinosa Engl.	Combretaceae	Multipurpose
Terminalia brownii Fresen	Combretaceae	Multipurpose

Preparation of cuttings

Cuttings were harvested from seedlings, managed juvenile stock plants or coppice shoots. Depending on the species, 1-to 4-node cuttings were used. These were usually about 50-60 mm long and with a leaf area of about 50 cm^2 (Whitcomb, 1983). In large leaved species, leaf areas were reduced by trimming prior to severance. The basal end of cuttings were dipped briefly in indole-3yl-butyric acid solutions (0.2-0.4% IBA in industrial methylated spirit) to a depth of about 2-5mm, and the alcohol then rapidly evaporated off in stream of cold air from a fan (American Association of Nurserymen, 2001). To minimize stress, the cuttings were inserted in the propagator as soon as they were dry. Alternatively, commercial auxin-based rooting powders "Strike" and "Seradix 2" (May & Baker Ltd) were used.

The non-mist propagator

The propagator design currently in use is based on the American Association of Nurserymen, 2001 modified by Morey et al., (1999) and now further modified so that it does not require daily watering. Basically, a wooden or metal frame is enclosed in clear polythene so that the base is water-tight. The frame also provides support for the enclosed volume of water. The polythene base of the propagator is covered in a thin layer of sand to prevent the polythene from being punctured by the large stones (6-10 cm), which arc placed on it LO a depth of 10-15 cm. These stones are then covered by successive layers of small stones (3-6cm) and gravel (0.5-1.0 cm) to a total depth of 20 cm. The gravel provides support for the rooting media, which is the uppermost layer, while the space between the stones, are filled with water. A length of hollow bamboo provides an open cylinder inserted into the medium and stones which is used both to observe the water level and to add water if necessary. The rest of the frame is covered tightly with a single piece of clear polythene, and a closely fitting lid is attached. Internal supports to the frame at the level of rooting medium also provide subdivisions allowing the independent use of different rooting media. A similar frame to that of the non-mist propagator, with roll up polythene sides, can be used.

Thepropagator environment

The test run in glasshouse, in which air temperatures were maintained at about 20°C temperatures in non-mist propagators rose to a mid-day peak of about 34°C during bright, sunny, mid-summer weather. This rise in temperature was associated with a decrease in relative humidity from about *95%* to about *75%* .

This represents a substantial increase in the saturation vapour pressure deficit (SVPD) of the air from 0.02 kPa to 1.37 kPa. An important decrease in relative humidity also occurred when the propagator was opened for five minutes at midday. In this instance relative humidity decreased by about 40-50% to glasshouse ambient within two minutes, representing an increase in evaporation rate of approximately x *4.5* (SVPD = 0.45 kPa to 2.08 kPa). Relative humidity increased rapidly again following closure of the lid. Decreases in air temperature to ambient were also associated with this period of opening. Subsequent gain in the temperature resulted from closing the propagator, but the response time for temperature was considerably slower than for relative humidity.

When the easy-to-root species *Nauclea diderrichii* was used for physiological studies in a non-mist propagator in Edinburgh University, it was found that cuttings had maximum rates of photosynthesis that were typical of intact plants, up to 6u mol CO_2 m_2s^1 at an irradiance of 1000u mol m^2s. However, the photosynthetic capacity of these cuttings was influenced by changes in the CO, concentration inside the prop-gator. In the middle of the day, the CO, concentration fell to 150 u *mol, mol*[1] while- by midnight it rose to 550 *p. mol mol*[1] reflecting daytime assimilation and night time respiration, respectively.

With regards to rooting media, the water-holding capacity of a fine-gravel (2-3mm diameter) medium was considerably increased, at the expense of the volume of the air spaces by the addition of rotted sawdust (50% by volume).

Tests of rooting

The studies using five tree species investigated the effects of four different rooting media (i) gravel, (ii) 50:50 gravel with saw dust, (iii) fine sand, and (iv) 50:50 fine sand and sawdust. Each medium was tested with cuttings dipped in a range of IBA concentrations (0, 0.05, 0.1, 0.2, 0.4 and 0.8%). There were, however, substantial differences between species with regard to rooting success on the different media. Single-node juvenile cuttings of all five species rooted well (70-95%) on their best medium. *Cordial alliodora* rooted best in fine sand, with or without sawdust,

The rooting of *Vochysia hondurensis* cuttings was detrimentally affected by the incorporation of sawdust into both gravel and fine sand. On the other hand, sawdust enhanced the rooting *of Eucalyptus degulpta* cuttings in both gravel and sand, while *Gmelina arborea* and *Albizia guachapele* rooted well in all media. Unlike these juvenile cuttings, mature cuttings from vigorous shoots in a heavily pruned crown of *G. arboreal* rooted much less well especially in pure gravel. In juvenile *G. arborea,* a comparison between cuttings set in fine sand under an intermittent mist propagator and the non-mist propagator showed better rooting in the non-mist propagator (American Association of Nurseryman, -2001).

C. *alliodora, A. guachapele* and *V. hondurensis* differed in their responses to the range of IBA concentrations (0 to 0.8%). Optimal concentrations would appear to be 0.4, 0.1 and 0.2% IBA respectively for the tree species. In addition C. *alliodora* did not root at all without applied auxins, while rooting in untreated *V. hondurensis* cuttings exceed 40%. Cutting of *A grachapele* were very responsive to all auxin treatments.

Studies using non-mist propagators in Edinburg and in Kenya, with tree species from semi-arid areas, investigated the relative merits of two commercially available, rooting powders (i) "Strike'" = 0.25% NAA. and (ii.» "Seradix 2" - 0.8% IRA. Seradix-treated cutting *of .Acacia tortillas, Terminalia spinosa* and *Terminalia brownii,* rooted better than those treated with Strike, while for *Prosopis julijlora,* rooting percentages were above 90% with both treatments.

Two-node cutting of *P. Juliflora* also rooted relatively easily without applied auxins, although IBA solutions (0.4-0.8 %) did hasten rooting and increased the numbers of roots formed. Long cuttings tended to root better than short ones, although there was no relationship between cutting length and its position of origin on the stock plant. Cuttings from lower on a stem generally rooted better than apical ones. An experiment comparing rooting under mist, with rooting in a non-mist propagator, clearly demonstrated the advantages of conditions without mist. A third treatment, in which a mist propagator was enclosed in polythene, resulted in even less rooting and higher mortalities due to rooting. When non-mist propagators have been used in Kenya to produce clonal material of *P. juliflora* for field experiments, the success rate, was greater than 75%.

Like those of *P. Juliflora,* cuttings of *A. tortilis* and *T. spinosa* are also much more easily rooted in the high humidity conditions of non-mist propagator. There is, however, very considerable clonal variation in rooting ability in these species, and cuttings of *T. brownii* have so far only been rooted with a low rate of success.

In Cameroon, the percentage rooting of cuttings of *Lovoa trichiloid.es* was relatively poor (c.40-50%), and, in addition, the results were frequently unexpected. For example, no beneficial effects on rooting were found following the application of NAA or a range of IBA concentrations (0-200 mg/cutting), either in terms of the percentage of cuttings rooted or the roots per cutting. Furthermore, the optimum leaf area seemed to be very high at about 200 cm^2 per cutting.

When the rooting of cutting of *Ricinodendron hudelotii* was tested both under mist and non-mist propagators with 'Seradix2', rooting by day 21 was best without mist (75% v 50% under mist).

The 'low-tech', non-mist propagators described here therefore seem to provide a very practical solution to the problem of how a wide range of tropical tree species can be propagated vegetatively. The advantage of these propagators seems to be particularly great for the dry-zone species, which can be very susceptible to rotting under mist.

While it is clear that the environment within the non-mist propagator fluctuates considerably over the day in response to variations in ambient temperature and incident radiation, it is also clear that by being enclosed and continuously moist, the cuttings are not subjected to the extremes of saturation and water stress that can occur when misting frequency is poorly matched to changes in the weather.

An improvement, which might further stabilize relative humidity in the non-mist propagator, would be to construct the lid in several sections. Then the whole propagator would have to be opened to access the cuttings. Ideally, the propagators are opened as little as possible and especially not during the heat of the day. When propagators are opened the cuttings should be sprayed frequently with a fine spray of water from a hand-held or knapsack sprayer. During bright sunny weather, shading is essential and spraying is highly desirable to prevent low relative humidity developing at mid-day. Attention to these points of detail is likely to be particularly important when propagators are used in hot, sunny and dry areas.

The reason why the rooting of cuttings of different species has slightly different requirements with respect to propagation media, and auxin concentration, is unknown. Studies are in progress at the Institute of Terrestrial Ecology (Leaky et al, 2001) and are aimed at the identification of fundamental principles determining rooting ability in a range of tropical tree species. In this regard it appears that stock plant light/nutrient interactions prior to severance are important affecting the subsequent capacity of uprooted cuttings to photosynthesis. This photosynthesis could also be limited to low daytime CO,, concentrations in the propagator. Thus it may be necessary in future to enhance CO., diffusion into the propagators.

The Biofertilizer Application to Forest Nurseries

The continuous use of chemical fertilizers, indiscriminate use of pesticides, poor crop, soil and water management, have not only decreased the productivity of soil, but also increasingly polluted the environment, causing serious health hazards to the entire biosphere. Also, in the light of recent global fertilizer shortage (both organic and inorganic) an intensive search for potential sources to supplement and/or substitute fertilizers through cheaper biological means has become imperative.

At this juncture, keeping all these in mind, use of bio-fertilizers throughout the world is becoming popular. 'Biofertilizer' literally means 'Life-fertilizer'. Biofertilizers are one among the numerous microorganisms, which live in soil (or) inside the roots of plants, and are beneficial to the plant as well as to the environment. Bio-fertilizers, which are widely used, are *Azospirillium, Phosphobacteria, Rhizobium* and *Azotobactor* .

Azospirillum, Rhizobium and *Azotobactor* are free living Nitrogen fixing bacteria and their population in untreated soil is usually very low; 102 to 104 in 0" to 6" depth of soil. When we apply these biofertilizers either directly to the soil or through seed coating, these multiply in enormous numbers and they absorb pure nitrogen from atmosphere and release required nitrogen in a sustained manner, so much so, it prevents the plants from nitrogen starvation. Among these three, *Rhizobium* is applied only to the leguminous plants, which forms root nodules. Thus all these biofertilizers improve the rate of seed germination, help in root proliferation, accelerate plant growth, and initiate early flowering and maturation. Phosphobacteria help in the mineralization of Tri-calcium-phosphophate and other nutrient complexes and thus increase the uptake by the plants. Out soils are naturally phosphorous deficient and by applying these micro-organisms, we can increase the uptake of phosphorous, which is essential in the development of root system.

For the first time in this country, 17 Lakhs forest tree seedlings have been inoculated with bio-fertilizers like Azospirillum, Phosphobacteria and Rhizobium at the rate of 1 gram per seedling by the Research wing of the Chennai Forest Department in their nurseries raised for supply of seedlings to catchment area planting under World Bank supported 'FamtFrrarru Agriculture Development project.

It is of interest to note that at Mukkombu Nursery, the average survival percentage of all the 10 species planted in the different treatments, when compared to the control, is more.

Results obtained at Mukkombu, Chennai, India.

Height increase over control in percentage:

Azospirillum	-	80.67%
Phosphobacteria	-	41.01%
Rhizobium	-	60.67%
Azos + Phos + Rhizo	-	55.25%

Biomass percentage increase over control:

Azospirillum	-	25.6%
Phosphobacteria	-	37.2%
Rhizobium	-	28.3%
Azos + Phos + Rhizo	-	32.4%

Thus, by applying these bio-fertilizers, we can overcome seedling mortality during the early years following planting.

8

Technique Used in Forest Seed Production

General Introduction

In agriculture, collecting seed from superior parent stock has been practiced for thousands of years. This technique results in higher yield and environmentally durable plant. This concept, however, has to gain widespread acceptance in forestry practice. To maximize yields and quality of trees for plantations, agro forestry, and other uses the following principles for collecting seed should be observed (Hibbered,2001).

The Seed Collection

In forestry, as in agriculture, the quality of offspring plants will be better if the seed used to produce them was collected from superior individuals, stands or orchards. Seed quality is measured in two ways. One, by the physical quality of the seed and secondly, by the desired physical traits of the resultant mature tree. The benefits of using better quality seeds, chosen from selected parent trees, are twofold: improved survival and greater economic returns.

Improved survival

The seeds from healthy, well formed trees provide greater assurance that resulting stock will have good form, survive and better resist stressed conditions due to marginal sites, frequent cutting, or harsh climates. In the adverse conditions typically weaken all but the strongest trees, making them vulnerable to insects, fungi, parasitic plants, and disease. It is very important because of the long-term resource investment and land and labour commitment in forestry, high survival rate of trees is important.

Farmers of Agriculture can recoup their losses after a poor season or even two. Trees, however, occupy a site for years. Therefore, any reduction in growth or quality from planting inferior stock represents a lost opportunity, which can be measured in time and capital, for as long as a tree occupies a particular site.

The Higher product yield

In the second benefit, economic return, the investment in seed selection will be more than compensated for by higher product yields over shorter rotations. Better yield result in more building materials, higher fruit production and quality, faster fuel wood growth and quicker and more prolific regrowth after lopping.

Simple measures are most effective

An effective seed selection can create success, even though sophisticated techniques may not be used. Any attention given to seed source will be a measurable improvement over current practices.

Besides careful seed handling practices must be included as part of the seed collection process. Such practices include transport, seed preparation, handling, cleaning, grading and seed testing.

Disregarding these practices will result in damaged or poorly developed seeds and allow the inclusion of unwanted or excessive amounts of foreign materials, which hinder germination and plant production.

The following sections in this chapter describe how to develop a system of proper seed collection and handling practices.

The Plannin & The Seed Collection Strategy

Too often in large planting programs the task of seed collection is an afterthought, typically left until the last minute and done hurriedly by unskilled or untrained labor while final nursery preparations are being made. Small-scale planters usually have access to leftover seeds from large programs. In either case, little consideration is given to seed quality, form or location of the parent stock.

Sometimes, to satisfy the required seed volume, nursery managers purchase bulked seed from villagers. However, villages tend to collect from the nearest trees or stands, which generally include immature, diseased, distressed and otherwise inferior seed.

And if seed needs to be collected from the crown, villagers will choose the easy-to-climb trees, which are typically small. Such trees would not normally be chosen as prime seed source candidates (Singh and Veerendra Singh 1982).

Also, in some species a single tree will produce a large amount of seed in some years. Such a harvest may satisfy the bulk requirement, but would be genetically dangerous because only one phenotype is represented. This technique causes problems for both large and small scale planters.

The Careful Planning Avoids Waste Time, Effort

By using untrained labor or collecting seed from a single tree, can be a critical mistake. Such practices often result in low survival, substandard trees, and a poor economic return for the time and work invested, for example nursery and site preparation costs.

In lied of an impromptu search or purchase from unknown sources, any seed collection, both small scale and large scale, should be a developed program of scheduled activities overseen by knowledgeable individuals. This is especially true for those involved in large scale planting programs; those involved in small-scale planting can often request advice from nearby large-scale planters.

An essential activities include selection of preferable parent trees; preparation of equipment including provisions for handling, transport, extraction, testing, grading and storage; recruitment and training of workers and seed sampling and quality monitoring. A timeline and work plan for the collection itself should be developed in conjunction with overall nursery operations. Most important of these is the selection of preferable parent stock. An able person who can distinguish between desired and inferior seed sources can accomplish this. This person, in turn, can supervise field seed collection during harvest.

Incorporating many of these activities often requires the level of funding available to large-scale operations. Never the less, even the small-scale planter should apply the principles as best possible whenever collecting seed (Hibbered, 1991).

Genetics Provides the Basis for Seed Selection

Why is it necessary to collect only from certain preferable trees? Tree improvement programs worldwide have shown that seed collected from trees with particular traits tends to produce trees with similar traits. One can correctly assume that if good parent trees are selected, then most of the resultant offspring will be of desirable quality. Yet despite this concept's proven value and wide understanding, it is* seldom applied in the field. To correct this shortcoming, simple parent tree selection guidelines should be developed. These would identify the desired traits and specify the parent trees that reflect those traits. Such traits might include straight trunks, fast growth, and better form.

These guidelines are normally met in national or company research programs, but following them will benefit all planters who are careful about the source of their seeds.

The traits in turn are chosen to reflect the ultimate desired use of the tree, including building materials, windbreaks, shade, fuel wood, food, fodder, or erosion control.

At the same time, these guidelines should not be made cumbersome. For the benefit of those individuals collecting seeds the guidelines could be a condensed set of simple criteria.

Indeed, immediate benefit can be derived from using even the most basic concepts such as selecting trees of good form and vigor and avoiding trees that are poorly developed, diseased, dying, or are insolated from others of the same species. These simple methods should be applied to large and small programs.

Collection Strategy Includes Area, Number and Spacing

While it is important to collect seed from the best parent trees it is also important to gather seed from several parent trees. Collecting seed from several parent trees assures diversity among seedlings being planted.

Whether the program is large or small, maintaining diversity is important and will help guard against pest epidemics.

To obtain adequate diversity, seed lots should be collected from a number of trees. In plantations and along roads more emphasis should be placed on collecting from the best trees. In all cases care should be taken to collect seed from several tree representatives of the best parent stock. Researchers have yet to determine an ideal number of trees per seed lot, i.e. those seeds that will be used in one planting. However, most agree that collection should not rely on only one or two individual trees. According to some sources, minimum of 15 to 25 parent trees per bulked seed source is preferred.

Furthermore, collection should cover a broad geography, including the environmental extremes at the edge of the range. Seeds selected from a narrow sampling of trees should be avoided as this will limit genetic diversity. Keeping a proper distance between selected parent trees in natural stands is also vital. During natural regeneration of some species seeds fall near the parent. At times inbreeding may occur, which will result in lower quality individuals. To reduce the chance of collecting a seed lot predominated by such trees a 100-meter distance between collection trees of the same species is recommended. This practice is especially critical to species that are looped or coppiced as a method of harvest and regeneration.

The temptation to forego a minimum number and spacing strategy is perhaps greatest during years of abundant seed production. Here large seed quantities are available from relatively few trees. At such times, an even greater effort should be made to ensure genetic variety.

Even better abundant seed years can be opportune. When properly cleaned and prepared, seeds of some species can be stored for several years. Stored volumes of good seed will help ensure supplies during production cycles. Where possible, all large-scale planting programs should have at least two years of seed supply in storage.

Conversely, leaner seed years present other challenges. During such years shortfalls will occur and the urge will be to collect for quantity without regard to quality. If at all possible, seed collection during lean production years should be minimized.

TPROPER SCHEDULING IS VITAL TO QUALITY

Aside from generic and site qualities, other considerations must be factored into the collection strategy.

For instance, only mature seed from ripened fruits should be gathered. Harvest schedules must account for the different times at which fruits from various species ripen. Depending on species and even different trees, seeds can be mature for a period of two to six months. To help determine when seeds are ready for collection periodic surveys of the selected stands is recommended. Survey should begin after flowering, as is the practice with cocoa. To prevent any false readings, specific sample trees should be marked and used throughout the process.

The surveys will also serve as an early indication of the season's expected seed volume and quality. In addition, field surveyors should look for early infestations of insects, disease or other seed predators.

With such information, the harvesting strategy and schedules can be adjusted to accommodate any foreseen hindrances.

Why Collect Seed From Several Provenances ?

Differences in how trees adapt to environment is often more important within a species than among species. There can be important differences in growth between seed sources from different geographical areas and environments. The areas and environments in which stands have developed through natural selection are called provenances. The growth, measured in height, in tropical hardwoods of the most vigorous provenances can be 30 to 50 percent greater than the growth in the least vigorous provenances (see figure 8.1). Unfortunately, there are very few tree species that have been adequately explored in their center of origin and studied in provenance and progeny tests on multiple sites, leaving much uncertainty about provenance which is intended to represent an entire species. Perhaps the most striking example of this comes from the use *of Acacia mangium* in Sabah, Malaysia where ultimately thousands of hectares were planted from seed collected from one tree. As a result, growth rates of the third generation were one-half those of the first generation.

A Field Strategy

How to select good trees

Parent tree selection in the fields is best done as a planned team effort coordinated by a trained person.

This selection of preferred trees serves to eliminate errors that would otherwise be produced by last minute collection efforts. In addition, a planned, timely program allow for the full use of available people and resources.

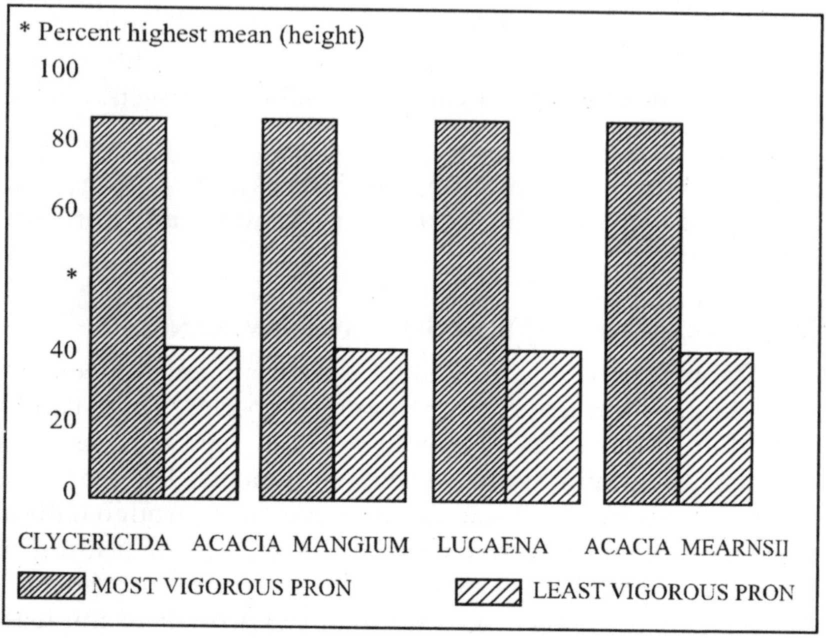

Diagram G1. Tree growths can be improved 30 to 50 percent using seeds from the most vigorous provenance.

Development of the collection strategy includes a series of routine screenings of potential parent stock. The series is broken into three progressive phases.

Phase one is done well before the onset of flowering. Give simple guidelines for site selection and appearance, general laborers search for appropriate stands. Locations of such are recorded, along with general descriptive information.

Phase two involves technical staff. Here, the identified stands are reviewed for their potential as a seed source. The review includes analysis of the site makeup including soils and topography; species composition, volume and spacing, and age, size and health of the trees.

Initial selections are made as to potential parent stock. These trees are marked for further study in phase three.

In phase three, trained professionals examine the marked trees for a final selection. The trees are analyzed for their physical characteristics, including their form, branching, growth rate, dominance, crown cover, health, and any clues as to their seed productive capabilities.

Before the final candidates are chosen, additional criteria are again considered. This includes the previously mentioned spacing preference of at least 100 meters between individuals of the same species. And, preference will be given to those trees located in the stand's perimeter.

Finally, selected trees are marked and their location recorded. Monitoring will increase as flowering begins and continue through the seed harvest period. Monitoring includes volume and quality of the seed as it matures in addition to any indications of infestations including insects and disease (Tewari, 1991).

The Seed handling and preparation

Seeds are rarely taken directly from the collection site for immediate planting in the nursery.

Instead, they must first undergo preparation that will ensure proper germination and help eliminate bad or damaged seed. Preparation includes all activities from collection to sowing, such as transport, sorting, extracting, cleaning, grading and testing.

In addition, seed harvesting and sowing periods are often months apart, which requires adequate storage containers. The containers must protect the seed stock from the weather and intrusive pests such as animals, insects or molds.

In some instances, proper temperature control may also be required, especially if seed is to be kept over a number of years.

Handling in the field

Once the fruits are picked or collected, they must be transported directly for seed processing. Any time lag in the field can cause dehydration or predation by various seed pests. The seeds must also be well protected from the weather to prevent excessive contact with moisture, which may prompt early germination.

In addition, because of the variation in harvested fruits, handling techniques must be flexible. Those in charge of seed storage must have adequate knowledge as to each species requirements and limitations during transport. For example, while the hard coated seeds of most leguminous species travel well, other seeds are more fragile and require greater handling care.

The nursery manager must have facilities readied for seed preparation at the time of collection.

Proper conditioning limits of seed degradation

Such facilities should be located at the nursery site, thus eliminating the need for further transport before sowing.

Seed extraction should occur as the fruits arrive from the field. Again, any lag between collection and preparation will cause the seed to degrade or spoil.

Depending on species equipment for both dry and moist extraction may be needed. Dry extraction requires appropriate sorting and drying areas, tumblers, and screens for sieving moist extraction calls for removing the fleshy outer coating of the seed, or the fruit, followed by drying and cleaning of any remaining foreign materials.

In either case, all foreign materials must be removed as these can result in poor germination or hamper storage ability.

Also the rate of drying and moisture content must be carefully monitored during. the drying process. Seeds dried too quickly may become damaged.

While small-scale planters may not have access to sophisticated equipment for drying and cleaning seeds, they should aim to remove as much foreign material as possible.

After extraction, the pure seed is then graded. Grading helps ensure a more homogenous seed set, which in turn will produce more uniform growth in the nursery.

Grading is done either by seed size, weight, or a combination of both. Actual seed size within the same species will vary due to a number of factors. The goal is to eliminate undersized, immature, or deformed seed. This usually can be done by hand if great quantities are not required. However, graders must be provided samples of the desired well-graded product to use as a guide (Anon, 1991).

In addition to grading, seeds are tested for germination rates, and are examined for phytosanitary concerns and species purity.

Using these factors, the forester must develop criteria for consistent seed certification. No matter how carefully done, collection efforts in the field will be nullified if inconsistent methods of grading and testing are used from one seed lot to the next.

Proper storage is essential

If the prepared seed is not used within a reasonable period of time, adequate storage must be provided.

As noted, most leguminous seeds store well in normal conditions provided they are properly cleaned, dried, and protected. Such prepared intact seed can be handled and stored without fear of damage.

However, if storage is required for several years, special facilities, such as cold storage may be needed for some species.

On the other hand if germination begins, different measures must be taken. Here the seed ceases dormancy and begins to use stored nutrients for survival. At this point, its shelf-life is drastically reduced.

Take note of growing sites

Aside from desired end-use traits, the tree must adapt to its growing site. In particular, it must withstand periodic harsh conditions, such as drought, and be able to grow in poor soils.

Physical traits of parent trees alone may yield few clues as to site adaptability. Here, environmental distinctions may help.

The collection staff should note even the slightest variation through out a species; range, topography soils, microclimate, associated vegetation, and man-caused factors such as animal grazing may all influence how a species conforms to its surroundings.

In overall location, a species typically grows best in middle of its range and fares less well at the edges.

However, seed lots should reflect all site variations, thus assuring that no one habitat is relied on too heavily.

But some cautions are needed when sampling for habitat, when associated plants are used as site indicator species, one must distinguish between natural conditions and manmade alterations. For example, continual grazing can heavily degrade a site, even to the point where it changes the vegetative makeup (Technical Bulletin, World Bank, 1991).

In addition, extreme dry or wet areas should be excluded if not within the normal bounds of a species' requirements.

During collection of particular stands, efforts should concentrate on trees that comprise the stand's perimeter. These individuals produce better quality seed.

Aside from natural areas, manmade stands should be reviewed carefully before selection as a seed source. For example, early reforestation efforts may have given little thought to seed collection, thus producing inferior parent trees to those found in natural stands. Or, the species planted on the site may not be well suited for the area. Whenever possible, historical data or records should be referred to for such stands.

A short list of collection tactics

- Collect seed lots 15 to 25 individuals that are spaced at least 100 meters distance from any other collection tree of the same species.

- Choose trees in vigorous health and avoid any that are diseased, suppressed, deformed, environmentally stressed, or in otherwise poor health.
- Collect from trees well formed and either dominant or co-dominant in the canopy.
- Avoid trees that are isolated from others of the same species.
- Harvest only mature seed from ripened fruits.
- To ensure genetic variation, collect fruits equally from all parts of the crown-top, sides, and bottom-as these parts may have been pollinated at varying times from different sources.
- Collect throughout a species normal habitat, noting variation in site.
- Man-made stands, including live fencing plantations, or windbreaks, should be carefully reviewed as to their establishment before being selected as a seed source.

Seedling Quality

An Quality versus quantity

A nursery manager's most important goal is to produce quality trees. Quality is more important than quantity. It is a common mistake in nurseries to concentrate on the total number of trees produced and

neglect their physical and genetic quality. It is better to produce a few-good trees, than many poor ones. Improving plant quality may mean that the farmer plants fewer trees, but the growth and survival of these trees will be superior.

Good plant quality is the basis for tree planting success. It is not worth a forester's effort to transport plants to the field, prepare an area, plant and maintain trees unless they are of good quality. A poor quality tree will always be one even if planted on a well-prepared, good site. In the field, each poor quality tree wastes space and resources leading to low site productivity. High quality trees have a higher survival rate and faster growth in the field than poor quality trees. Fast growth allows a tree to out compete with weeds and reduce the initial labor costs of establishment. Fast growth also enables a farmer to harvest wood or tree products, sooner, increasing the return on the fanner's investment. We are producing trees for people's livelihood; they depend on having high quality tree (Kumaravel, 1993).

Seedling quality has two main aspects. The first is the genetic quality or the sources of the seed; the second component of seedling quality is its physical condition when it leaves the nursery. Improving genetic quality of seedling requires a long-term strategy of seed selection, while improving the physical quality can be accomplished in just one or two seasons.

Quality of Seed source

The farmers select only the best animals for breeding; animals that are small and sickly do not produce good offspring. Similarly, tanners use only the superior crops that have high yields and are resistant to disease for the next year's seed. These same principles should apply to trees. The characteristics of the parent trees can greatly influence the characteristics of the seedling. The seed can determine whether the tree will grow well or poorly. Studies from around the world have shown that good seed improves survival, timber and fruit quality, and shortened rotation or harvest times,. Because trees take longer to mature than crops or animals, thus making tree planting a long-term investment of labor and land, it is even more important to select only high quality seed.

The desired characteristics of the parent trees will vary depending on whether the trees are for wood, fodder, fruit, or medicine. A good nursery practice is to consult farmers as well as forestry technicians when selecting the seed sources. Farmers often know additional traits, which make their trees more valuable. It may be difficult to find some of the trees with the best traits because these are often the first to be cut down. Conserving some of the best trees within the community will ensure future supply of seed.

Some desirable parent tree characteristics are:

- Healthy trees with a large, well developed crown
- For timber trees, a long, straight trunk with few branches
- Wood quality, such as high density, or straightness of the grain
- For fodder trees, pulatability and digestibility of foliage (leaves that animals like to eat and are easily converted into energy)
- For fruit trees, low branching may be desired for easier fruit harvest
- Fruit quality, such as sweetness or ability to be transported with minimum damage
- Fast growth rate
- Low susceptibility to (or ability to quickly recover from) disease or insect attack.

A Good nursery practice is to select the parent trees well in advance, and plan a way to ensure sufficient seed is collected. Permanently marking the trees as seed sources may help ensure that they will not be cut down. O-y mature seed from ripened fruit should be harvested. A Good nursery practice for each species trees is to collect seed from at least 30 parent trees that are at least 100 meters apart. If you buy the seed, find out how many trees were used. Using a large number of seed sources increases the gene pool or genetic diversity of the seedling. Genes are the codes of information from the parent trees that determine how the progeny or offspring trees will grow. By using seed from many different trees the pobability of the offspring trees having good characteristics increases and ensures, that the trees can better adapt to environmental changes. A poor, but unfortunately very common nursery practice is to collect seed from just one, two or three easy-to-climb trees close to the nursery. If the seed is bad and does not germinate, the nursery crop could be lost. Collecting from just a few trees is also dangerous because it results in low genetic diversity. Trees with low genetic diversity are often more susceptible as a group to disease, or unable to adapt to changing environmental conditions such as drought. If an area is planted with trees from very few sources, in the future, our ability to choose the best seed sources and improve the characteristics of the trees is very limited.

Use seeds from an area as similar as possible to the area where you are planting. For example, seed from a mountainous region should only be planted in a mountainous region and seed originating from the lowlands will grow best in lowland conditions. For long-term genetic improvement though, seed from each individual tree is kept separately and tested in field experiments. The best trees are then selected to serve as seed sources for the nursery.

The principles of tree domestication are similar to those used in agriculture; maximize the quality of tree products, maximize tree growth rates, ensure the adaptability of species to the planting site, and maximize resistance to disease and pests. Selecting the best seed sources and managing the trees under optimal conditions achieve.

Physical quality of Seeding

Not a single characteristic determines seedling quality. Seedling quality is a combination of height, diameter, plant nutrition, health, root size and shape. Together, these characteristics determine how well the plant will establish itself in the field, and they affect the rate of survival. Height alone is often not a good predictor of how a plant will grow in the field. A good nursery practice is to judge seedling quality by several traits.

Many of these traits act together and influence one another. The goal of producing the best seedling is to optimize these traits while producing specifically what is needed for a particular site. You will need lo talk to foresters and farmers to find out the most important desired traits. For example, plants for dry, rocky soils need to be short and be produced in small containers, whereas plants for flooded sites (or) active pastures may need to be quite large.

Quality tree seedlings have the following characteristics:

- They are healthy, vigorously growing and free of diseases.
- They have a robust and woody (lignified) single stem free of deformities.
- Their stem is sturdy and has a large root collar diameter.
- Their crown is symmetrical and dense.
- They have a root system that is free of deformities.
- They have a dense root system with many fine, fibrous hairs with white root tips.
- They have a 'balance' between shoot and root mass.
- Their leaves have a healthy, dark green color.
- They are accustomed to short periods without water.
- They are accustomed to full sunlight.

The following examples demonstrate how these traits enable the plants to be more resistant to transport and planting stress, and how they improve seedling growth and survival.

How quality traits work together

Sturdiness' is measured by the relationship between height and diameter. If two trees have the same height, and one has a larger diameter stem, then the latter is sturdier. A sturdy stem is less susceptible to transport and planting damage.

Stem diameter is often related to root size. Plants with large diameter stems often have large root systems. Diameter is a better predictor of root size than plant height.

A 'balanced' seedling has a small to medium shoot system and a large root system. The shoot loses water through the leaves and the roots compensate for this loss by absorbing water and nutrients. "Unbalanced plants have too many leaves and too few roots. Balance refers to the mass or dry weight of the shoot and root, NOT to their length.

Assessing seedling quality

We do not need special equipment and it does not take long to survey seedling quality. When the seedlings are about 15 cm tall, choose at least 20 plants from each species for inspection. It is important tosample randomly, that is, not to pick only the biggest plants or only plants from one bed. Try to sample plants from all parts of the nursery and from each bed in the nursery. Choose one or two plants from each end from the middle of each bed. Examine the plants thoroughly. 16 out of 20 plants (80%) should have the quality characteristics listed above. If fewer than 16 trees are of good quality, repeat this quality assessment at least once more about 1 month before the plants are taken to the planting site so that necessary improvements can be made.

A good nursery practice is to sacrifice a few plants to improve the quality of the total nursery production. One of the best ways of examining quality is to cut open the container to observe the root system of several plants; of course, these plants, must then be thrown away. Do not plant these seedlings afterwards, because when inspecting the root system, the fine root hairs will get damaged, and the plants will either die or become badly stunted.

Root deformities - the hidden curse

Root deformities below the soil line are the hidden curse in seedling production. They retard growth, cause the plant to lean or even fall over, and can result in the plant's death. Root deformities do not correct themselves over time, in fact, they become more acute as the tree grows.

The main root should be as straight as a carrot, or if there is no main tap root, the many smaller roots should branch out without any pattern or strong bends. If the roots are in knots or coiled, they will eventually strangle the tree or they may die, attracting insects and fungi that will damage the tree. There are two types of root deformities:

(A) Roots deformities caused by poor pricking out from the germination bed into the container

The deformities are generally within the first 10cm under the surface of the soil or at about a finger's length. Often, seedlings are squeezed into holes that are too short for the roots system. Roots are stuffed forcefully into bags or, while placing a seedling into a hole, the end of the root remains curled upwards. Because roots always want to grow downwards, they will bend back and grow into a 'knee' or even a complete loop. Nursery customers can check for these deformities by following the stem down with their finger. These plants should be refused because they will never grow well in the field.

(B) Roots deformities caused by the bag

Smooth plastic bags cause the principal root to coil or spiral along the walls or at the bottom of the bag or pot. This inevitably happens when plants are left in the nursery too long. However, it can also happen to plants that are only a few centimeters tall. Plants commonly develop roots before they begin shoot growth. So even plants with small shoots may have long roots that are coiled at the bottom of the bag. These roots should be cut off immediately before planting.

Table G1

Problem	Solution
Roots knotted or twisted from poor pricking out	Discard plants immediately. Next time, direct sow, or follow procedures for correct pricking out.
Roots coiled at the bottom of the bag	Cut roots with machete or prunes, before planting. Remove plants in good time from the nursery. Use root trainers.
Roots penetrating into the soil beneath bags	Lift bags and prune roots frequently. Remove the plants in good time from the nursery. Use root trainers on frames above the ground.

Multiple plants per bag	Remove extra seedling early, before they become very big
Plants with multiple stems	Discard-the cause may be poor genetic quality.
Diseases or insects	Isolate or burn any affected plants. Develop a plan for pest management.
Yellow or white leaves, or leaves with dark green'or purple veins and light spots in between	Fertilize plants or use a richer substrate.
Large variation in plant sizes among plants sown at the same time	Check for patterns within the beds and throughout the nursery that can be associated with uneven shade or watering.
Plants grow slowly	Adjust light (try more or less), fertilizer or use a better substrate,

Source: Wightman 2001

A good nursery practice is to regularly survey seedling quality to correct problems through appropriate nursery management. Some problems, and suggested solutions, are given in Table 8.1.

Keep the best, ditch the rest

In every population of trees, there will always be some plants of good quality and some of bad quality. On an average, 20-30% (and often even 50%) of the trees will be of poor quality. Thus, the nursery should always produce 20-30% more plants than needed to meet a given target. Nursery managers must accept that it is quite normal to throw away plants. Unfortunately, many do not; many people do not want to see their hard work thrown away, or they believe that the plants still might have a chance. These ideas are false.

A good nursery practice is to discard poor quality trees as soon as you detect them. They waste space and resources in the nursery, and can be a source of infection. The process of removing poor quality seedlings is known as 'culling'. It is a greater waste of hard work and money to maintain trees of poor quality in the nursery and in the field, than to throw away poor quality trees in the nursery.

Variable plant growth in the nursery is carried over to the field. A weak plant will never catch up with others that were strong when planted out. A poor, but unfortunately common, nursery practice is to select the best trees for planting, but then leave the bad ones in the nursery. These are then given to the next unsuspecting customer. This is very unfair to the customers. Only the best quality trees should leave the nursery, the others should be thrown away at that very moment. A plant that is not considered good quality does not suddenly become better when the best ones have been planted.

Another poor, but unfortunately common, nursery practice is to leave plants in the nursery from one production year to another. These are usually the left-over plants that no one wanted, but the next planting season they are overgrown and have severe root deformities. If plants grow very slowly during the year and remain in the nursery more than one season, throw these ones out, and try adding more compost to your substrate, or allowing your plants more sun.

Seed Pre-treatments and Seed Storage

If seeds are enclosed in a fleshy fruit, remove as much as possible of the flesh with a knife, wash off the rest under water, and plant the seeds immediately. Fruit flesh attracts insects and fungi, which may damage the seed, so it's important to plant the seed as soon as you can. For seeds in a seed pod, such as a bean, let the pods split open naturally by laying them in a semi-shaded place. Similarly, for other fruits with a woody coating, drying them in semi-shaded or gently cracking should open them and let you collect the seed. Some sun is good, but be careful not to overheat, which might kill the seeds.

A good nursery practice is to pre-treat the seeds, if they take more than a week to germinate. You will save time and resources by pre-treating seeds. Faster production time in the nursery means lower nursery costs and may also allow the farmers to plant the seedlings during the best planting times. If you are unsure how long the seed will take to germinate, sow some before you sow all of the seeds to find out. If most of the seed germinates well with the first treatment method, then that is probably sufficient.

The seeds contain two parts; an outer protective coat, and the inner embryo that develops into the plant. Germination begins when water penetrates the seed coat and the seed swells. When seeds swell, plant them immediately. Discard seeds that float; they probably have air pockets caused by insects or dead embryos. Always use 2-5 parts of water for each part of seed (e.g. two to five tablespoons of water for one tablespoon of seed). Change the water every 12 hours to remove chemicals that may also slow down germination.

Do some tests to see the best orientation for sowing the seeds. This is especially important if the shoots or roots are twisted when they emerge from the seed, which is common in mahogany and mango. In southern Mexico, the common way to sow mahogany seedling is with the wing pointing up. But in one nursery, the manager noticed that the best way to plant was with the wing pointed down. In Puerto Rico, it is recommended by a forest research station that seed be planted flat.

Four basic seed pre-treatments

(i) *Cold water treatment:* Soak seeds in cold water for at least 12 and up to 48 hours. You can also try soaking the seeds in water during the day, and leaving them to dry at night.

(ii) *Hot water treatment:* Boil water in large pot, remove from heat and cool for 10 minutes. Add seed to water and let soak for up to two days.

(iii) *Boiling Water treatment:* Boil water in a large pot, remove from heat, add seed and leave for two minutes. Pour off the hot water and replace with cold water. Soak the seed for up to two days.

(iv) *Mechanical:* It is important not to damage the embryo or the part of the seed where the root will emerge (usually an indented place or a pointed area). Nick the seed with a knife, crack it with a stick, or scratch its surface on a concrete floor, or with sandpaper. For small seed, place in a jar lined with sandpaper facing the inside of the jar, and shake vigorously. It is not necessary to completely remove the tough outer seed coat just crack it so that water can enter the seed and trigger germination.

A good nursery practice is to sow the seed immediately after collecting, or store correctly if necessary. You should plan ahead and have containers filled, so that seed can be sown without delay. If you must store seed before sowing, make sure it is kept well ventilated in a cloth or mesh sack, or in clay pots. If a refrigerator is available, you can use plastic bags or containers.

Only clean, dry seed should be stored. Hang the sacks from a pole so that air can circulate all around them. Keep the seed in a cool dry place, out of direct sunlight. A fungicide may help reduce fungus attack. Mothballs or other strong-smelling materials like cedar wood may also help reduce insect attack. Check seed regularly for signs of decay, insect or rodent damage.

The length of time that seed can be stored varies greatly between species. Some seed can be stored for many months, while other seeds lose their viability. If the seed has been stored for more than one month, a good nursery practice is to test the germination rate to see if the seed is still good. A poor, but unfortunately common nursery practice is to observe.

Seed Technology 135 sow old seed in many containers, without knowing if it will germinate-you might be wasting time, labor and resources. Sow 100 seeds in a shady germination bed and water normally. Count the number of seeds that germinate, and this will tell you how many you need to put into each container to achieve one plant per pot. If for example, only 25 out of 100 seeds germinate, sow four in each container (Wightman, 1999).

The Seed Germination

The most important factor for good germination is constant humidity around the seed. Seeds can be covered with sieved soil, sand rice hulls or pine needles, to keep them moist. A light material allows the tender roots and shoots to emerge without hindrance. The substrate should be moist, but not soggy; excess water should drain off. It is often recommended that you plant large seeds deeply and small seeds close to the surface. While this is usually sensible, if the substrate dries out quickly, or water is not always available, you will need to plant the seeds deeper. If the water pressure in the hose is very high, it may wash the seeds out, so again, plant the seed deeper. A good nursery practice is to carefully control the light, water and shade conditions during initial seedling growth. Shade helps retain humidity and prevents new leaves from scorching. This can be achieved in a seedbed as well as directly in container

The substrate should not receive additional fertilizer because this could increase the risk of diseases such as damping-off. Germinating seedlings generally receive all nutrients they require from the cotyledons (or in the case of palms, from the first leaf) that are formed inside of the seed. Thus, substrate-like sand, which does not contain fertilizer, is generally a good germinating medium. Damping-off is common with small seedlings and is recognizable as either decayed seed or. more frequently, decay around the stem at the soil line. The leaves of the seeding droop as though they need water, although the substrate is wet, then the stem appears "pinched" and brown near the base. Eventually, the seedling falls over at the soil line.

Damping-off is caused by fungi which can be present on the surface of the seed in the substrate. Seed can be sterilized by immersing it for 30 minutes in a 10% solution of Clorox (1 tablespoon of bleach plus 9 tablespoons water). Hydrogen peroxide, a common antiseptic available at the pharmacy, is also an effective seed sterilant. Its advantages are that it is less toxic than bleach, and it can increase germination because it softens the seed coat allowing water and oxygen to enter more easily. Seed is soaked directly in the antiseptic for up to four hours. With both bleach and hydrogen peroxide, some experimentation may be necessary to find the most effective solution strength and soaking time. A 30 minutes soak with hot (just too hot to touch comfortably - definitely not boiling) water also works well on seed. The substrate can be sterilized by wetting the area well with water then covering the area with a black plastic sheet for several days while it stands in abundant sunshine. The heat kills bacteria, fungi, and weed seeds. You can sterilize sand by washing it several times until the wash water appears perfectly clear.

When deciding whether sterilization is necessary, first determine whether seedling growth is being harmed by something living in the soil by carefully examining the type of damage to the plant. Is the damage at the soil line or at the roots? If yes, then sterilization may be necessary. Test the different methods to determine if seedling growth really improves. A good nursery practice is to carefully recognize the type of damage on the seedling, and control the specific problems with the right method for that particular pest. Remember that sterilization of the substrate can also kill beneficial fungi, bacteria and insects. Certain bacteria and fungi are necessary to break down organic matter or help plants to absorb nutrients better. Many insects too are useful because they eat other insects that eat plants.

Some of the common problems with seedbeds are that:

- Pricking out almost always results in root deformities
- When plants are left in the seed bed too long (which often happens), they produce long roots that are easily damaged or twisted when pricked out
- Many seedlings suffer a 'shock' when they are transplanted from seed beds. Their growth is slowed for I -4 weeks before they begin to show visible new growth
- Pricking out is often done by staff with little nursery seed Technology 137 experience, and even people with experience often prick out incorrectly, and
- Paying people by the number of plants pricked out in one day gives no incentive to prick out correctly.

When to direct sow seed

Careful control of water, light and nutrients is required during germination and during the first few weeks of seedling's growth. Directly sowing the seeds in the container saves time, labor and money, because the extra step of preparing a seedbed and transplanting is eliminated. Even if it takes little longer to plant small seed directly in the containers or if they have to be moved and resown, this is easier and cheaper than pricking out. Direct sowing allows undisturbed seedling growth and this reduces stress for the seedling. In a well-managed nursery, seedlings can almost always be directly sown! (Wightman, 1999).

When you are directly sowing, follow these good nursery practices

- Use only fresh and ripe seed
- Pre-treat seed, if necessary, to speed up germination
- Prepare containers and shade in advance
- Mix small seed with sand or rice hulls, or use a bottle with a screen top (like a salt shaker, but larger), to make dispersing it easier
- Test seed for viability before sowing. If less than 70% germinate, sow more seed again in the bag. Throw away any extra seedlings in each pot. This is a small price to pay to avoid root deformities.

When to use seed beds

Depending on the conditions in your nursery, including the tree species, number of plants produced, and labor availability, a combination of direct sowing and use of seedbeds may be your best way of operating. The use of seedbeds is NOT recommended simply in order to ensure every container has a seedling. Seedbeds can be used:

- To select seedlings of uniform size and development for transplanting. Only plants of the same age should be compared when judging plant quality. When seed germination is highly erratic, seedlings should be transplanted in groups of the same age. Seedlings should be pricked out in groups of even ages so that later they will not be compared with plants of a different age class.

- When seed is old, or when the germination is low or unknown. Use as seedbed to test seed viability before filling too many containers and wasting resources.
- If seed does not store well (that is, if it is 'recalcitrant').
- If containers are not available, or not filled in time to use; Seedbeds can be used until the containers are ready.

Types of seed beds

There are two types of seeds beds, temporary and permanent. Temporary seed beds include any tilled bed of soil. Alternatively, seeds can be germinated between bags of jute or sheets of newspaper that are kept constantly wet, but not soggy. The advantage of using newspaper or a similar material as opposed to planting in a bed of sand or soil is that you can easily check the stage of seed germination and prick the seedling out as the taproot emerges. It is not necessary to wait until the seedling grows 15 cm and produces its first true leaf, and in fact that's not a good idea, because the older, better-developed seedlings are more susceptible to damage when pricked out.

Plastic bags filled with a sand and soil mixture are common. The bag can be closed initially to provide shade and a humid microenvironment. Small seed lots can be germinated, and the substrate can be easily loosened to remove the seedlings. Another advantage of these bags is that they can be moved directly to the rows of containers where the seedlings will be planted.

Permanent seed beds are commonly constructed with cement blocks at 1 m or more heights and 1.2 m width for working comfort and efficiency. They are filled starting at the bottom with stones, then gravel, then soil or sand. A common practice is to prepare wooden boxes with a fine wire mesh bottom and fill them with sand. The boxes are placed on table in a greenhouse or under a roof. The tables are also covered with fine wire mesh to allow good drainage and aeration. Frequent watering is necessary because the sand does not retain much water. Creating the proper germination conditions in containers, and pricking out correctly, are more important than constructing tidy, permanent germination beds (Hibbered, 1991).

How to prick out correctly

Even when pricking out is done very carefully as described below; it is still very hard to do it correctly, especially when thousands of plants must be transplanted. Only those who are trained in correctly pricking out should attempt it. Unfortunately, many people who have pricked out in the past have done it incorrectly. Close supervision and revision of the plants is necessary in order to avoid root deformities. Always open a few containers after a few days to check that it has been done correctly.

Good pricking out practices

- Throw away any seedling that appears sick or deformed.
- Transplant when the tap root emerges or seedlings are still small (5 cm), before secondary roots are formed.
- Water the bags well, one night before you prick them out, so that water penetrates to the bottom of the seed bed.
- Ensure that the area where the transplanted seedlings will be kept is well shaded before you begin pricking out.
- Water the seedlings 24 hours before, and one hour before, pricking put.
- On days with strong sunshine, prick out in the early morning or late afternoon.
- Use a shovel or stick to gently loosen the soil around the seedling.
- Remove seedlings by grasping their cotyledons or lower leaves do no lift them out by the stem.

- Put seedlings in water as soon as you take them from the germination bed.
- Prepare planting holes with a stick and ensure they are sufficiently wide and deep.
- Clip long or very branched roots to ensure they are pointeddownwards.140 Techniques in Forestry
- Gently pull the seedling upward after placing it in the hole, to straighten out roots.
- Pack the soil against the roots, starting at the bottom of the hole.
- Water the plants immediately after transplanting, and again when they wilt.

Poor but unfortunately common pricking out practices

- Waiting until plants are large and have long roots.
- Pricking out plants into dry soil and then watering them.
- Constructing shade after pricking out is done.
- Pricking out in direct, hot sunlight.
- Transplanting damaged seedlings.
- Removing seedlings by grasping the stem, as this may permanently damage the flow of water.
- Carrying seedlings in your hand or on a plate without water.
- Preparing the holes with a finger-the hole will usually be too small.
- Allowing roots to bend upwards when inserting them into the hole.
- Leaving air pockets around the roots - the plant will die.

Summary of sowing and seedling germination

Whenever possible, direct sow of seed into containers, will avoid root deformities.-Controlling the germinating environment and correctly pricking out are more important than constructing fancy seed beds.

Good nursery practices

- Sow seed as soon as possible after harvest.
- Pre-treat seed, if it takes more than a week to germinate.
- Test old seed before planting to know if it will still germinate.
- Carefully control the light, water and shade during seedling growth.
- Sow seed directly in the container.

Poor, but unfortunately common nursery practices

- Expending a lot of effort to build fancy seed beds.
- Careless pricking out resulting in damaged plants with root deformations.
- Leaving seedlings in the germination bed too long.

9

Technique Used in Bamboo Propagation

The bmboo plantations can be raised by the following methods:

A) Direct sowing of seeds
B) By culm with roots and rhizome
C) By the stock with roots and rhizome
D) By rhizome with roots
E) By offset planting
F) By the cutting of culm
G) By branch cutting
H) By tissue culture

Sowing of Seeds Direct

Characters of Seed

The possibilities of raising bamboo plantations from seeds are not always practical because of the unusually long seeding cycles. Except in the case of *Melocanna baccifera,* seeds of most of the commercially important species of bamboo resemble grains of paddy (or) wheat and are light in weight. Seeds for direct and nursery sowing should be obtained direct from the forest. The seed may be collected either on the ground previously cleared (or) on a cloth spread around the culm, by shaking the culms. The best time for collecting the bamboo seeds is between the months of February and July consequent to the flowering. The number of seeds per kilogram varies from 75,000 to 1,05,000. The bamboo seeds remain viable for short periods, with the initial viability is being good, the seeds rapidly lose germination vigour. The longevity of seeds varies from species to species but they are generally viable upto 1 -2 months, although the period can be increased under controlled storage conditions. Seeds of *Bambusa tulda* stored in a desiccator over silica gel maintained their viability even after 18 months (Banik, 1987). By storage of seeds under suitable temperature and moisture, the longevity of *Dendrocalamus strictus* seeds could be extended upto 34 months (Gupta and Sood, 1978). The viability of seeds varied from 30-35 days for *Bambusa tulda,* 55 days for *Dendrocalamus longispathus,* 65 days for *Bambusa arundinacea.*

The Seeds *of Thyrostachys siamensis* can be stored for about 27 months under controlled conditions (Anantachote, 1985). Seed viability can be extended by reducing the initial moisture content before storing, which helps in maintaining viability up to 9 months. Seed samples stored at room temperature lost their viability within 21 months, in comparison with seeds stored for 27 months under low temperature such as in the cold room (2-4 °C) and deep freezer (-5 °C), and could maintain a high percentage of viability (89.2 - 92.5) and were able to germinate within 3- 4 months (Ramagarangasi, 1988).

The Seeds should be given pretreatment before sowing, i.e. normally soaking in water for 48 hours. Bamboo germinates generally within 5-10 days of sowing and the seedlings attain 1-leaf stage in 7 days. Percentage of germination is higher when seeds are sown directly in polythene bags rather than in nursery beds.

Direct sowing method

For direct sowing, the soil should be dug upto a depth of about 10-15 cm, (or) deeper in case of poor soils, either in line 3 m apart, (or) in patches 3m x 3m (or) 4.5 x 4.5m apart and cleaned of all weeds. Seeds, after having been soaked for 48 hours before sowing are dibbled in beds at the plantation site and slightly mulched. The resultant plant is kept clean weeded for the first 2 (or) 3 seasons. 1 .25 kilogram of seed per hectare will suffice for widely spaced patch sowing and about 1 1.5 kg for line sowing. Though direct sowings are an easy method, they are generally neither practicable nor desirable owing to shortage of seed, the necessity of continuous weedings till plants are well established and the liability of destruction by seed eating birds and rodents.

Transplanting method

As direct sowings in lines are often liable to fail due to relatively (i) uneven germination, (ii) slow growth of resultant plants, and (iii) exposure to animal damage etc., the more usual procedure is to raise plants in a nursery bed for transplanting entire plants. For this purpose the nursery site should be located well within the planting region and near an adequate supply of water.

At first the seedlings look like grass, whippy shoots appearing after a few months. Great care is required in planting out bamboo seedlings, which are liable to wither if the roots are exposed to the sun. Planting should be done as quickly as possible, before commencement of monsoon season. Transplanted seedlings are irrigated two hours regularly both in the morning and evening. Weeds are removed from the plantation area as and when required. After one year, the plantation should be adequately irrigated at 15 days interval. Care must be taken to avoid water logging and protect against rodents, grazing and browsing animals.

By Culm With Roots and Rhizome Method

In this method, the new young ones sprouted in the same year (or) the previous year should be used. Although large-size culms are preferable able and the small-size ones also can be used. Leaving the branches with several nodes, the culm with branches of upper part is removed. The rhizome of the culm should be yellowish, young, vigorous and bearing good buds. Old rhizomes should not be used, for they are lacking the buds that develop into culms. The length of rhizome is 40-60 cm in length with about ten nodes and buds. Fibrous roots on both culm and rhizome must be left. The rhizome should be cut carefully with a saw; cutting with a hatchet will cause damages to the buds.

By Stock With Roots and Rhizome Method

In this case the culm is branched and its length is about 30 cm. The procedure is the same as in the above method.

By Rhizome With Roots Method

Rhizome without culms *are* cut 50-60 cm long with about 10-15 nodes and their roots. The 2-3 year old rhizomes with roots are most satisfactory, but over 5 year old rhizomes are entirely unsuitable. This method is recommended for transplanting to distant places. In this case, they are wrapped with sphagnum moss and covered with vinyl sheets after the soil is washed away. Usually, they are first laid in the nursery bed 20 cm deep and covered with soil. They are transplanted next spring, as new culms begin to grow. This method is applicable to monopodial type.

By Offset Planting Method

This is the good method of vegetative propagation of clump -forming bamboo species in which success has been achieved. One year old culms which are ready to sprout within the year are cut off with short rhizomes. The culm should be about 40-50 cm in length. The success of this method depends in part on the vitality of the rhizome stock used and the time of the year when it is planted. If the rhizomes are taken from young healthy stock and planted immediately at the break of rain success can be expected. But if the rhizomes are taken from old stock and planted much before the rains, complete failure may result. The weather of the year must also be an important factor.

By Cutting of Culm Method

This method, however, has a limited use because of the limited availability of the planting material. Rooted culm cuttings can be successful with most bamboo species. The age of the culm and the period when these cuttings are put out for rooting are important considerations. The two year old culms and spring season have been found to give best results.

However, a method of vegetative propagation using culm cuttings has been standardized at the Kerala Forest Research Institute (KFRI). This involves treatment of culm cuttings with growth regulating substances for including root formation. The method was tested for *(Bambusa anmdinacea, B. balcooa, B. polymorpha, B. ventricosa, B. vulgaris, Dendrocalamus brandisii, D. hamiltonii, Ochlandra scriptoria* and *O. travancorica)* 12 species of economically important bamboos and was found successful. Large scale multiplication of superior varieties is possible by this method. When out planted, plants raised from cuttings develop to clumps much faster than seedlings. This portion describes the practical steps involved in propagation of bamboo through culm cuttings.

Collection of material and preparation of cuttings

An extract 2 to 3 years old culms from healthy culms by cutting just above the first node during March-April. Trim the leaves and side branches. Care should be taken not to injure the auxiliary buds while removing leaves and branches. Transport the culms to nursery site as quick as possible. Maximum care should be taken to prevent drying. This can be done either by wrapping in moist gunny bags (or) embedding in boxes containing moist sawdust. Two-noded cuttings (a cutting with two nodes leaving about 5-7 cm on either side of the nodes) must be prepared using a sharp knife (or) saw. For bamboos with thin walls use of a saw is advised to avoid splitting of the cut ends. Make an opening (about 2 cm in length and 1 cm in width) or drill two holes (about 7 mm diameter) in the centre of the inter node.

Treatment for root induction in cuttings Method

A 10 gram of NAA (1-Napthalene acetic acid) BDH, Loba Chem (or *E.* Merck), is dissolved in 250 ml of ethyl alcohol

(90%) in a container by stirring the solution gently and add water to make up to 100 liters. The solution is mixed thoroughly by stirring. The final concentration o1 NAA will be 100 mg/1 of water (or) equivalent to 100 ppm. This solution is sufficient to treat 1000 cuttings. About 100 ml of the solution is poured to the culm cavity. To avoid spillage, use a wash bottle to pour the solution through the drilled holes. The holes were closed by rapping a drying witr a polythene strip (60 cm x 6 cm). Ensure that the polythene wrapping is tight so that the solution does not leak out. The cuttings were kept horizontally with the opening facing upwards. For bamboos with narrow culm cavity like *Dendrocalamus strictus* the treatment is given by dipping the basal portion of the culm cutting in solution of NAA for 24 hours After extraction, the culm cuttings should be treated with NAA as quickly as possible (preferably the same day). If the planting site is far away and there is unavailable delay for planting, the treated cuttings can b presented up to three days by keeping in moist sawdust.

A Preparation of nursery and planting technique

Raised nursery beds of 10m X 1m is prepared and filled with a mixture of soil and sand (3:1). One weak prior to planting, drench the nursery bed separately with the insecticides, Aldrin and the fungicide, Bavistin to prevent termite and fungal attack. For each bed, use 40 litres of 0.015 % (9.1). Aldrin prepared by adding 0.5 ml of Aldrex 30 EC per litre of water and 30 litre of 0.05 % (9.1). Bavistin is prepared by adding 1 g of bavistin 50 WP in one litre of water. The treated cuttings were placed horizontally (the opening facing upwards) across tfoe nursery bed. About 50 - 60 cuttings may be conveniently planted on a raised nursery bed of 10 m x 1 m. The cuttings were covered with a thin layer (2-3 cm) of soil.

Nursery management technique

Nursery beds must be provided with a thatch to protect the cuttings from direct sunlight, which may be removed at the onset of monsoon. The beds should be watered regularly in the morning and evening with 30 to 40 litres of water per bed at each watering. Care should be taken to avoid water logging. The sprouts after one month are treated with 0.01 % a.i of Bavistin to avoid fungal attack. If necessary, _ farmyard manure may be applied to increase the vigour of the sprouts.

Transplanting method

Rooted cuttings can be uprooted and transplanted to the field during (June - July) (after about 4 months). If cuttings are sprouted and rooted at both the nodes, cut carefully at the middle of the cutting to get two plants.

Precautions observed

The collection of culms from clumps which are likely to flower in the near future is to be avoided, because new clumps developed from such cuttings will also flower and perish along with the mother clump. The year in which the clumps are likely to flower can be found out by checking the previous flowering record from the area and the flowering cycle of the species.

By branch cuttings method

In this method, branch cuttings from the bottom, middle and top portions of the healthy culms are taken from the clump. With a sharp knife, the branches are cut at their basal portion, the leaves and side branches are removed and cuttings with two internodes are prepared. The cuttings are immediately transferred to the nursery. A natural rooting medium (mixture) is prepared by mixing rice husk, dried powdered cow dung, charcoal and sandy soil in equal proportions. The mixture is filled in black polythene bags of 45 cm height. The cuttings are placed in a thatch to protect from direct light. They are regularly watered in the morning and evening. The cuttings sprout within 20-30 days after planting and root development, can also be observed within this period. Rooting efficiency is better in the samples taken from the bottom region. The branch sprouts start to develop from that region. The rooted cuttings are transferred to the soil in August. July and August are suitable season for the vegetable multiplication of bamboo using branch cuttings. The method has the following advantages:

(i) It is simple and can be adopted by the farmers.
(ii) There is no need of costly chemicals like hormones.
(iii) A large number of propa gules can be obtained without cutting the culm.

By Tissue Culture and Macroproliferation Method

Large scale production of bamboo seedlings for raising bamboo plantation is hampered by the long vegetative cycle and death of the flowered clumps, coupled with low speed viability. Although planting stock can be produced by vegetative propagation, the synchronous flowering of the vegetatively propagated clumps and parent clumps leads to the death of the entire plantation. At this juncture tissue culture assumes great importance to replenish the acute shortage of planting stock as the cultures can be maintained in juvenile phase for long durations.

Technique of Tissue culture of Bamboos

The concept of tissue culture is borne out of the fact that, every cell in a plant is capable of regenerating into an entire plant, thus making it possible to regenerate innumerable number of plants from a single bud. A number of protocols are published for multiplication of bamboo. All the available protocols have been tested and viable protocol for mass multiplication and transfer to field have developed been at plant Biotechnology Division of Forest Genetics and Tree Breeding, Coimbatore. Protocols for multiplication have been developed for *Dendrocalamus strictus, Bambusa arundinacea, Dendrocalamus asper* (edible bamboo), *Phyllostachys edulis Dendrocalamus giganteus* and *Bambusa polymorpha.*

The Collection of bud materials

The single nodal buds from identified phonotypes (or) axillury buds from seedlings/young plants can be taken as the basic material for culture. The time of collection of bud materials from the mature culm is critical. They are collected at the time when the buds start swelling. In case seedlings material is used as explant, nodal, portion from one month old seedling are preferred for multiplication.

Sterilisation of explants

The buds are washed thoroughly with running water and sterilised with 0.1 % Sodium hypochlorite for 5 minutes, washed with sterile distilled water and with 0.1 % Mercuric chloride for 5 minutes and all traces of the sterilant removed by rinsing with sterile distilled water. In the case of nodal explants from mature culms more rigorous sterilisation is advocated.

The entire operation is carried out in the laminar flow chamber to maintain aseptic condition. The explants are initiated in Murashige and Skoog medium supplemented with 0.5 mg/ltr Benzyl Amino Purine. The cultures are maintained at a photoperiod of 12 hrs dark and 12 hrs light with cod fluorescent lamps (100µ mol m^{-2} s^{-1}).

Preparation of media

The buds are initiated in a medium containing essential nutrients for growth. Basic bamboo culture medium ideally suited for multiplication is given below.

(i) Bamboo culture Medium (Stationary liquid culture) developed at Plant Biotechnology division of Institute of Forest Genetics and Tree Breeding

Compounds	mg/ltr
Ammonium nitrate	1650.00
Potassium nitrate	1900.00
Calcium chloride	440.00
Magnesium sulphate	370.00
Potassium dihydrogen phosphate	170.00
Potassium iodide	0.83
Boric acid	6.20
Manganese sulphate	22.30
Zinc sulphate	8.60

Sodium molybdate	0.25
Copper sulphate	0.03
Cobalt chloride	0.03
Ferrous sulphate	27.84
Ethylene diamine tetra acetic acid	37.30
Myo-inositol	100.00
Nicotinic acid	0.50
Pyridoxine	0.50
Thiamine HC1	0.10
Glyane	2.00
Sucrose Benzyl	30000.00
Aminopurine	0.50

The above chemicals are weighed accurately and dissolved in 1 liter double distilled water. The pH of the medium is adjusted to 5.0. The medium is then dispensed into jars covered with suitable enclosures and sterilized in the autoclave for 20 mts at 15 Ibs pressure. The media may need modification depending on the physiological status of mother plants at the time of bud collection. For explants derived from seeds modification is not necessary.

(ii) Sub culture

Within one months of culturing, the buds start producing multiple shoots. Regular sub culturing is done every 20-25 days m the same medium. Within 20 days, about 20-30 shoots are produced. These are furjher sub cultured with three shoots per flask.

Rooting and out planting Method

Much of the shoots on prolonged culture are rooted in the same medium. 90% of the plants root successfully in MS half strength medium supplemented with 0.1 mg/ltr Indole Butyric acid. Micro rhizomes are also induced under in vitro condition so that plantlets become hardy. This also ensures high survival rate when transplanted in polybags. The rooted plants with microrhizome are then transferred to a medium containing sand, soil, vermiculite in the ratio of 1:1:1 and kept in poly globule for hardening for a period of two weeks. If required, during this period 0.1% Bavistin is sprayed to avoid fungal attack. Subsequently, the polyglobutes are opened and the plantlets kept in shade house for a period of one month and then transferred to the field. While the plantlets are kept in the shade house, Hoagland's solution is fed to the plants to enable them to rapidly gain height and dry matter. Within a month the hardened plants are transferred to the field.

Transplanting technique

The seedlings are planted at 6m x 6m spacing with a total of 250 seedlings per ha. The transplanted seedlings are irrigated two hours regularly both in the morning and evening. Weeds are removed from the plantation area as and when required. After one year, the plantation is adequately irrigated at 15 days interval. Care should be taken to avoid water logging and protected against rodents grazing and browsing animals.

Production of culms

All transplanted seedlings begin to produce rhizomes. From these rhizomes lateral culms will emerge which are the chief causatives for the total biomass yield. The number of culms developing from the rhizome totally will constitute a clump. The productivity of bamboo is assessed by the number of new culms produced annually. At a given site the production of new culms is mostly dependent on the culms of the previous year, the degree of congestion and the clump age.

A Macro proliferation

In order to increase the planting stock before transfer to the field, macro proliferation is practiced. Four to five months old plants in the nursery developed with the above procedure has good rhizome system. These plantlets in the nursery develop five (or) six tillers at four to five months period. These tillers are separated along with a small piece of rhizome and roots. These propagules attain the size of field plantable saplings within four months (or) they can be further multiplied through macro proliferation.

Thus by this technology a large number of identified planting stock can be made available. Using this technology one bud can be made to produce three hundred plants in three months which can further be multiplied on a larger scale. Essentially the approach is same for all the species of bamboo, viz., *Dendrocalamus strictus, Bam bus a anmdinacea, Dendrocalamus asper* (edible bamboo), *Phyllostachys edulis, Dendrocalamus giganteys* and *Bambusa polymorpha.* Varying the cytokinin concentration and effectively subculturing within a period of 20 - 25 days would help in continuous production of propagules with high multiplication frequency. The propagule production is limited by manpower availability for subculturing frequency and space availability in the tissue culture laboratory and rapidity with which it can be transferred to the field. Several thousands of prop gules can be produced by careful planning and anticipating the limitations. It is advisable to obtain the total planting stock requirement, at least a year in advance and work - back the multiplication strategy.

Scheme for mass multiplication of Bamboos

Seed from identified phenotype

↓

Aseptic germination

Identified Phonotype ↓

Explant (singal node)

↓

Explant in culture medium

↓

Multiple shoot formation

↓

Continuous culture for multiplication

↓

Rhizome induction

↓

Root induction

↓

Rooted plant kept for Acclimatization
↓
Propagation through Macro proliferation
Nursery - Macro proliferation ↓
Field

SCHEDULE TIME FRAME FOR PROPAGULE PRODUCTION OF BAMBOO WITH APPROXIMATE MULTIPLICATION RATE

Single auxiliary bud
One month ↓
Multiplication upto 300 shoots in 3 months
50% ↓
Continuous culture
↓
Rhizome induction
↓
Root induction
↓
Macro proliferation 1X5

For future
multiplication - 50%
4 months ↓
Nursery
50% ↓
Field

Season of planting

The time of planting is a most decisive factor in the ultimate success of the plantation. Planting work must be undertaken immediately after the first showers of the monsoon. If it is rhizome planting, when the buds on rhizomes slightly show a sign of swelling, it is the best time for the selection of good shoots. Caution must be taken in transportation of such rhizomes for prevention of injury of the buds. Suitable irrigation recommended when bamboos are planted in mountain area. The new shoot that has sprouted in the same year may be planted very carefully with rhizomes during the rainy season in and after June. In the tropical regions, the rainy season from spring to summer is the optimum season for planting.

Number under planting scheme

When the soil is fertile or the species is a type of large - size culms, plantation is carried out less densely than in the case of poor soil or of the species developing culms of small size. Generally, the number to be planted is 300 - 500 per ha (space 4-6 m). So, before planting the soil analysis is to be carried out, to supply the nutrient status in the soil.

Method of planting of Bamboo

Instructions for planting are as follows;

(i) The junctions of culms and rhizomes should not be injured.

(ii) The bamboos should be planted, after being dug up as soon as possible. For the transportation to a distant place, rhizomes are cut with roots but without the culms and then wrapped with sphagnum moss after the soil has been rinsed away.

(iii) After planting, they are irrigated and fertilized. The direct contact of fertilizers with the bamboo roots should be avoided. The fertilizer is applied in the following proportion: per ha. 200 kg of 15:15:15 N.P.K fertilizer is sufficient for increasing the production of culms.

(iv) On a sloped land, rhizomes are planted in such a way that the buds come out on the upper side of the slope.

A Guidelines for Bamboo Management

A taditionally, bamboo has been referred to as minor forest produce. However, the status of bamboo has changed considerably and it is rapidly emerging as an important plant group in many forestry programmes. During the past few decades, several Asian countries have embarked on large - scale cultivation of economically important species. The important guidelines for raising and management of bamboo are discussed here.

Although the raising of bamboo plantation by vegetative methods is found successful, the availability of required propagating material is scarce. Seeds possess short period of viability and the procurement of seeds is also difficult. So, the seedlings obtained from tissue culture methods are used. The large sized polythene bags 20cm x 30cm are filled with clean soil, sand and manure in the ratio 2:1:1. About 15 to 20 plants are planted in each polythene bag. The nursery is adequately irrigated and care should be taken to avoid over saturation. Seedlings are kept in polythene bags for one month. Nursery beds were provided with a coconut thatch to protect the seedlings from direct sunlight.

Each polythene bag contains 15 to 20 seedlings. The number of rhizomes may vary from 4 to 5. These set of rhizomes are not singled out or sorted out, but the entire thing in one polythene bag is planted in the field.

The seedlings are removed from their polythene bags carefully and transplanted to 45cmx45cmx45cm pits in the fields at a spacing of 6mx6m. One week prior to planting, the pits are treated with 0.01% aldrex and 0.05% bavistin to prevent termite and fungal attack respectively. The planting should be completed before the monsoon fully sets in.

To provide better initial growing environment to the seedlings, the upper halves of the pits are filled up with a mixture of 25 gram 15:15:15 N.P.K. fertilizer per pit, while lower halves are filled up with the original soil. Protection against damage by rodents, grazing and browsing animals is to be carried out. The watering is done, but over watering is avoided. The transplanted seedlings produce rhizomes, which develop new rhizomes that produce lateral culms, which are the chief causatives for the total biomass yield. The number of culms developed from the rhizome totally constitute a clump.

At a given site the production of new culms is dependent on the culms of the previous year, the degree of congestion and the clump age. The average annual recruitment of culm generally increased year after year. The culms produced during first year are shorter in length and smaller in diameter, but culms from subsequent years will be longer in length and bigger in diameter.

The rhizome development is not peripheral as generally believed. Rhizomes may develop is any direction and the culms may appear anywhere in the clump provided overhead light is available for the emergence of the culm. It is only in congested clumps that the new culms appear to grow at the periphery. The new culms can even be seen in the middle of the clump.

New bamboo culms are produced each year. Older culms should therefore be harvested every year. Culms older than two years take little part in the growth of the new culms. Retention of culms older than two years is therefore not needed for the production of new culms. Culms older than three years are fully mature are desired for use. The following rules should therefore be applied for getting high productivity of the desired quality of culms:

1. All bamboo culms three years older should be harvested.
2. Harvesting should preferably be done each year.
3. Where congestion has already set in the congested culms must be removed even if it leaves only the current year culms.
4. Where the young culms are twisted from the top they should be cut, so that new culms grow freely.
5. No gelling operations should be carried out from July to October.
6. Mounding (or) heaping earth around the bamboo culms should be carried out after every felling of culms.
7. Bamboo clump should in no case be clear felled. The clear felled clump generally degenerates into a bushy form.

The above mentioned siliviculture and management practices can be used as guidelines for taking up large scale plantations on suitable site.

A protocol for Micropropagation from seedling explain

(i) Collect fresh viable seeds.

(ii) Place these seeds in water, and treat them with 5-10% (W/V) mild detergent solution.

(iii) Shake vigorously and leave it under running water for one hour.

(iv) Wash the explants so as to free from Treatment solution.

(v) Sterilize the plants in 0.1% (W/V) HgCl, under the laminar Airflow Chamber for 5-10 minutes.

(vi) Wash the seeds in sterile distilled water for 2-3 times.

(vii) Inoculate the seeds in a freshly prepared semi solid White's basal medium containing 2 % sucrose and 0.42% agar.

(viii) After 25 days transfer the well-developed shoots to the shoot multiplication medium (1/2 strength MS liquid medium containing 2% sucrose, BA 0.5mg/1 and Kn 0.5 mg/1 supplemented with Coconut water 200 ml/1).

(ix) Place one shoot per flask on orbital shaker at 90 rpm.

(x) Add fresh media for every 15 days interval.

(xi) After 30 days transfer the elongated shoots (5-7 cm) into ½ MS basal medium containing 2% sucrose, 0.42 % agar supplemented with IBA 0.25 mg/1 for rooting.

(xii) Place the tubes in dark for 48 hr in a culture room

(xiii) Select the rooted explants from the medium and transfer to a sterile beaker.

(xiv) Wash the explants with sterile distilled water and dry with a blotting paper. I-5-.

(xv) Transfer it to a plastic cup containing sterile Vermiculate: Soil (1:1) mixture.

(xvi) Cover the cup with plastic paper and keep it in a growth chamber or culture room.

(*xvii*) Maintain the cultures at 27-30 °C, 85-90 % RH, and 16/8 photoperiod.

(xviii) Maintain for 2-3 weeks.

(xix) Remove the plastic bag.

(xx) Transfer the plants to fresh pot containing normal soil and keep it in a mist chamber.

(xxi) Plants may take 4-5 weeks to acclimate.

A protocol for Micropropagation for mature tree explant

(i) Collect fresh nodal explants from a healthy tree.

(ii) Cut the nodal explants into single shoot segments.

(iii) Treat with 5-10% (W/V) Teepol solution.

(iv) Shake vigorously and leave it under running water for one hour.

(v) Sterilize the plants in 95% ethyl alcohol for 3 minutes followed by 0. 1 % (W/V) HgCl, under the laminar Airflow Chamber for 5-10 minutes and wash the explants in sterile distilled water 2-3 times.

(vi) Trim the nodal explants to 0.5 to 1 cm.

(vii) Inoculate the nodal segments in a freshly prepared White's basal.medium containing 2 % sucrose and 0.42% agar.

(viii) After 25 days remove the well developed shoots from the medium and transfer to 1/2 strength MS liquid medium containing 2% sucrose, BA 0.5mg/l and Kn 0.5 mg/1 supplemented with Coconut water 200 ml/1 for shoot multiplication.

(ix) Place one shoot per flask on an orbital shaker at 90 rpm. J-0.

(x) After 30 days transfer the elongated shoots (5-7 cm) into a 1/2 MS basal medium containing 2% sucrose, 0.42 % agar supplemented with IBA 0.25 mg/1 for rooting.

(xi) Place the tubes in dark for 48 hr in a culture room.

(xii) Select the rooted explants from the medium and transfer to a sterile beaker.

(xiii) Wash the explants with sterile distilled water.

(xiv) Transfer it to a plastic cup containing sterile Vermiculate: Soil (1:1) mixture.

(xv) Cover the cup with plastic paper and keep it in a growth chamber or culture room.

(xvi) Maintain the cultures at 27-30 °C, 85-90 % RH, and 16/8 photoperiod.

(xvii) Maintain for 2-3 weeks.

(xviii) Add 1/10 strength MS macro elements stock at 3-5 day intervals.

(xix) Remove the plastic bag.

(xx) Transfer the plants to fresh pot containing normal soil and keep it in a mist chamber.

(xxi) Plants may take 4-5 weeks to acclimate.

10

Restoration Eco- Methodology

General Introduction

Reclamation of mined dumps and mine spoils are the major concerns in protecting the environment (Barber 1984). Magnesite mine spoils, limestone mine spoils and quartz dumps are some of the problematic soils existing in Tamil Nadu. The Institute of Forest Research, Coimbatore, has developed packages for afforestation of these soils. The technology package includes physical and biological rehabilitation.

Physical Rehabilitation

This constitutes the major part of reclamation process of mine spoil.

- Analysis of physical, chemical and biological properties of mine spoil.
- Removing and storing the topsoil before mining for future afforestation.
- Leveling and shaping of the mine dump with convenient aspect and terraces, contours for planting.

Biological Rehabilitation

Rehabilitation of the minespoil can be made after studying the serial stages of various pioneer species, selection of suitable tree species and soil amendment techniques which include trenching around the plant, mulching with organic materials and cover crops, application of biofertilizers and inorganic fertilizers (Wittwer, 1974 ; Power and Papendick, 1985).

The Establishment of herbaceous vegetation

The locally available seeds and tillers, which will improve the site through their root system, may develop certain pioneer species like grasses and legumes and the resultant organic matter gradually changes the texture of soil for further succession.

The Species selection

Species which are found locally and those that are capable of colonizing degraded areas should be preferred. The criteria for species selection may be listed as follows:

- Adopted to grow, spread and reproduce under severe condition.
- Provide quick green cover and soil binding.
- Leguminous species to fix atmospheric nitrogen and ameliorate the soil by addition of organic matter through plant litter.
- Fruit yielding trees to abode birds and other forms of wildlife and also encourage soil fauna.
- Plantation species with good economic, social and aesthetic value for local population.
- Fast growing indigenous and exotic species to accelerate plant succession.
- Control of erosion and stabilization of mine dump.
- Fire hardy, unbrowsable and stress tolerant.

Best performed species for Magnesite minespoil,Limestone minespoil and quartz sand dump

Miignesite minespoil	Limestone minespoil	Quartz sand dump
Acacia nilotica	*Eucalyptus tereticornis*	*Casurina equisetifolia*
Casuarina equisetifolia	*Acacia holosericea*	*Acacia auriculiformis*
C. junghniana	*A. auriculiformis*	*A.nilotica*
Ailanthus excelsa	*A. mellifera*	*Azadirachta indica*
Morinda tintoria	*A. nilotica*	
Ficus spp.	*Morinda tintoria*	
	Azadirachla indie a	

Species suitable for stabilizing the slopes of minedumps

Vinca rosea	*Tephrosia purpurea*	*Cassia auriculata*
Indigofera sumatrana	*Stylosanthes hamata*	*Agave sisalana*
Leucaena leucocephala	*Commiphora berryi*	*Cenchrm ciliaris*
Cymbopogan	*Jatropha curcas*	Local grasses etc.

Importance of tree species for the restoration of mine dump areas

Tree species	Importance
Acacia auriculiformis	Nitrogen - fixing tree, fuel
Acacia holosericea	Nitrogen - fixing tree, fuel
Acacia nilotica	Nitrogen - fixing tree, fuel
Ailanthus excelasa	Fast growing, drought tolerant
Alangium salvifoliwn	Fruit yielding
Annona squamosa	Fruit yielding
Azadirachta indica	Fruit yielding, multipurpose
Bambusa bamboo	Soil binder, drought tolerant
Casuarina spp.	Evergreen, fast growing and Nitrogen fixing tree, poli, fule
Emblica officinalis	Fruit yield, drought tolerant, attract birds, soil enriching, medicinal
Eucalyptus terticornis	Evergreen, fast growing, drought tolerant
Ficus benghalensis	Fruit yielding, attract birds, soil enriching
Ficus religiosa	Fruit yielding, attract birds, soil enriching

Holoptelia intergrifolia	Secondary timber, root suckers, drought tolerant
Morinda tinctoria	Fruit yielding, drought tolerant, secondary softwood, natural regeneration
Pithecollobium dulce	Fruit yielding, drought tolerant, attract birds & animals, natural regeneration
Santalum album	Fruit yielding, Sandal wood, attract bird & animals, natural regeneration
Tecoma stans	Attractive flowers, natural regeneration
Syzygium cumini	Fruit yielding, drought tolerant
Pterocarpus santalinus	Drought tolerant, Nitrogen fixing, timber
Wrightia tinctoria	Drought tolerant

Mulching

Mulches such as agricultural refuse, sugarcane refuse, sewage sludge, straw, coir pith and jalsakthi can be applied around the seedlings in trenches, terraces and slopes which will alter the microclimate of the surface and help to conserve moisture during the establishment of plants.

Application of Biofertilizers

Biofertilizers develop symbiotic association with plants which supply a protected niche to the microbes for growth and in return, the plants receive a continuous supply of atmospheric nitrogen and are capable of solubilising certain inorganic phosphates into utilizable forms through their enzyme action. Inoculating with suitable improved biofertilizer combination which is tolerant to heavy metal concentrations is one of the important techniques (Kumaravelu, 1993).

Biofertilizer dosage

Rhizobium	-	5 g/ plant
Azospirillum	-	5 g/plant
Phosphobacteria	-	5 g/plant
VAM	-	5 g/plant
Pisolithus tinctorius	-	5 g/plant

Specificity of Biofertilizers for Tree Species

Leguminous species	-	*Rhizobium*
Wide range of tree species	-	*Azospirillum,* VAM fungi and Phosphobacteria
Acacia & Eucalyptus spp.	-	*Pisolithus tinctorius*
Casuarina spp.	-	*Frankia*

Fertilizer application

Apply slow release fertilizers containing Nitrogen, Phosphorus and Potash during rainy season for establishing the vegetation in these areas and restore the fertility of the land.

The Forest Research Institute, Coimbatore has given technical consultancy in afforestation of floatation reject dump (Quartz sand) to ACC cement factory at Madukkarai, Tamil Nadu. This is a field level consultancy in which combination of species with suitable soil amendments were recommended. This demonstration technology is being extended in the Quartz mine dumps by the ACC.

Bio-fertilizer Application for Growth Enhancement in Nurseries

The application of bio-fertilizers in nurseries is gaining importance in the recent years while raising quality planting stock. The bio-fertilizers not only boost the growth of seedlings but also ensure better survival and good health (Kumaravelu, 1993; Jaya balan 1993). The results of the experiment carried out at IFGTB are:

Delonix regia performed well when treated with *Phosphobacterium* and *Azospirillum* showing 200% and 58% increase in height respectively, over control.

Cassia siamea when treated with *Phosphobacterium* and *Azospirillum* showed 102% increase in height over control in both the cases and showed 21% increase when treated with *Rhizobium.*

In young teak plantation, application *of Azospirillum* and *Pliosphobacterium* increased the height by 29% and girth by 9.5%.

The inoculation of *Frankia and Azospirillum* was found to be highly effective in *Casuarina equisetifolia.*

Ceiba pentandra responded well to *Phosphobacterium* and showed 58% increase in height over control.

Albizia lebbeck responded favorably to *Azospirillum* and *Phosphobacterium* inoculations showing 16-33% increase in height over control.

Triple inoculation of VAM, *Rhizobium* and *Phosphobacterium* in *Acacia nilotica* was found to be most promising in terms of growth.

In case of *Acacia nilotica* and *Albizia lebbeack, Glomus fasciculatwn* species was found to be the best performing VAM fungus, while *Glomus intraradices* was the best for neem. Farmers, NGOs and State Forest Department, are adopting these packages.

11

Technique Used in Plant Analysis

1. Mesurments of Fresh / Dry Weight of Plants

After growth of specific time duration, plants are collected and weigh these plants either whole or cut them into different plant parts like root, shoot, leaves and fruits etc. as a fresh matter. The weight of these fresh matter production indicate the fresh weight of plants.

After taking the fresh weight this matter is put in an oven at 80°C temperature for 72 hours. After that take weight again, this weight indicates dry weight of plants.

2. Measurments of Growth Parameters

Growth performance, such as, Relative growth rate (RGR), Net assimilation rate (NAR) and Leaf area ratio. (LAR) can be determined as suggested by Blackman (1968). Different growth parameters are determined by the following formula –

(a) Relative Growth Rate (RGR)

RGR is the increase in dry weight per unit day weight per unit time.

$$RGR(g/g/day) = \frac{Log\, w_2 - Log\, w_1}{T_2 - T_1}$$

(b) Net Assimilation Rate (NAR)

NAR is the increase in plant dry weight per unit leaf area in an unit time.

$$NAR(g/cm^2/day)=\frac{(w_2-w_1)(Log_e A_2-Log_e A_1)}{(T_2-T_1)(A_2-A_1)}$$

(c) Leaf Ana Ratio (LAR)

$$LAR(cm^2 g/)=\frac{(A_2-A_1)(Log_e w_2-Log_e w_1)}{(Log_e A_2-Log_e A_1)(w_2-w_1)}$$

Where, Ti and T2 are the time of first and second harvesting. wi and W2 are total plant dry weight of time Ti and T2 Ai and A2 are the leaf area at time T1 and T2

3. Measurments of Leaf Area and Dry Weight Ratio

The area of the leaf is plotted and measured, after that the same leaves are kept at 80°C for 48 hours till a constant weight is obtained. The leaf area dry weight ratio is calculated as follows :

$$Leaf\ Area\ Dry\ Weight\ Ratio=\frac{Total\ Leaf\ Area\,(cm^2)}{Dry\ Weight\ of\ leaf\,(g)}$$

4. Yield Parameter

On harvesting the crop, 3 or 5 plant replicates are taken for per plant yield attributes viz. number of pods or cobs per plant, number of seeds per pod or cob, number of seeds per plant seed weight and 1000 seed weight. Grain yield and harvest index are calculated on meter square area.

Harvest index is the ratio of grain yield per unit of biomass.

5. Pti (Percent Thermoinstability), Phsi (Plant Height Stress Index), Dmsi (Dry Matter Stress Index), S (Drought Susceptibility Index), Ys (Yield Stability Ratio)

These all parameters are calculated using the following formulae

(i) Bausloma and Schapaugh, 1984

$$PTI = 1 - \frac{1-(T_1/T_2)}{1-(C_1-C_2)} X100$$

Where, T_1/T_2 = Conductivity ratio under rainfed condition (E2)

C_1/C_2 = Conductivity ratio under irrigated condition (E1)

$$PHSI = \frac{Plant\ height\ under E_2}{Plant\ height\ under\ E_1} X100$$

$$DMSI = \frac{Total\ dry\ matter\ produced\ under\ E2}{Total\ dry\ matter\ produced\ under\ E1} X100$$

(ii) Fischer and Maurer, 1978

$$S = 1 - \frac{Yd}{Yp} / D$$

Where, Yd and Yd are the mean yield of the genotype under £2 and Ei conditions respectively and D is the drought intensity.

$$D = 1 - \frac{Mean\ Yd\ of\ all\ the\ genotype\ under\ E2}{Mean\ Yd\ of\ all\ the\ genotype\ under\ E1} X100$$

(iii) Lewis, 1954

$$YS = \frac{Grain\ yield\ under\ E2}{Grain\ yield\ under\ E1} X100$$

6. Measurments of Moisture Content

The moisture content of the plant materials (leaf, seeds, flowers, etc.) is determined, 'on fresh weight basis. Three replicates of plant materials are taken and fresh weight is determined. These materials are then kept for drying at 80°C in oven for 72 h. and then their dry weight is determined. The moisture content is determined as follows : –

$$(\%)\, Moisture = \frac{Plant\ fresh\, weight - Plant\ dry\, weight}{Plant\ fresh\, weight} X\, 100$$

7. Measurments of Relatfvt Water Content

It is measured following Fletcher et al. (1988). Turgid weight is found by placing the cut ends of three 1 g batches of leaves in small beakers containing water. The beakers are enclosed in a humid atmosphere at room temperature (25°C) for 6 hours. The submerged ends of the leaves are blotted rapidly and thoroughly before weighing. The leaves are then transferred to an oven at 120°C for 48 hours and their dry weight is determined. The mean RWC value for the 3 replicates is calculated from the formula –

$$RWC = \frac{(Fresh\, weight - Dry\, weight)}{(Turgid\, weight - Dry\, weight)} X\, 100$$

8. Measurment of Water Potential

Measurement by Dew-Point Microvoltmeter

Since the introduction of this Pressure-Volume (PV) analyzer has proven to be valuable tool in the evalutary plant water relation. It's main advantage is that a large number of important water relations parameter can be obtained by a single procedure. For example, the osmotic adjustment of plants to water deficit or salt stress can easily be reduced by comparing the osmotic potential at full turgor and at zero turgor during a drought cycle or after subjecting the plants to salt stress. Furthermore, the bulk modulus of elasticity and the water storage capacity of the tissue investigated can be calculated and their dependency on changes in relative leaf water deficit modelled with a pressure-volume analysis a Hofter diagram can be constructed.

Apparatus required : L-51 Psychrometer/hygrometer (Wescor Le Logan USA).

Modification in Instrument : Instrument is slightly altered as recommended by Shackel (1984).

The commercially available aluminium block was well insulated with plastic form and covered with aluminium foil to improve refleehan. Additionally, about 5 cm of the electrical lead is insulted by covering it with the same form in a PVC tube. To simply weighing during the PV measurement. The wire connecting the in situ hygrometer to the micro volt meter is cut about ten cm from the sensor and fitted with the connector to allow the reattachment to the receiver of the wire. Dew-Point hygrometer is calibrated with NaOl solution of known concentration.

Method adapted : At least two hours before the experiment is started, for the *in situ* observation, hygrometer is placed into growth chamber to ensure temperature equilibration. According to the HR-33T instruelean mameal, the cooling coefficient ^{A}v is at for drug, sealed hygrometer. Prior to the final installation the leaf is inserted into the leaf clamp. After making the shape of the aluminium specimen holder with small strips of taps, the leaf is removed. If the position of the hygrometer sensor the cutiles is gently separated away with emery paper (No. 600) to increase the conductance for water vapour exchange. The leaf is singed with distilled water and carefully blotted dry. A thin band of a mixture of Lavolin/ beeswax (50 : 50) is then applied around the seeling face of the hygrometer sensor. The sensor is then attached to the leaf in the clamp by gently pressing it against the leaf surface. Inspite of abrading the cuticle, initial water vapour equilibration generally required up to three hours as determined by Psychrometer measurement. The Dew-Point hygrometer remained attached for several days without causing obvious changes to the leaf.

The Dew-Point hygrometric determination of water potential is interrupted by a I'sychrometric measurement. The leaf is carefully cut off from the twig, the lead is disconnected from the microvolt meter and the leaf is weighed on an electrical balance. It is then put back into the chamber, reconnected to the microvolt meter and another hygrometer measurement is made. After an additional Psychrometric determination, while the leaf is allowed to transpire, the resulting changes in the leaf water potential would be monitored. This measuring cycle is repeated upto 20 times, until leaf water potential is for beyond the turgor loss point, which lasted more than 7 hours.

Finally the leaf is removed from the holder, cleaned and leaf area is determined by leaf area meter and keep in ovan for dry matter determination.

Preparation of Data Sheet

Dew-Point hygrometer readings

Date :

Name of Species :

Initial weight of leaf:

Weight of leaf after experiment:

Weight of whole arrangement + Leaf:

Weight of arrangement of leaf:

Leafarea :

Table

Sensor No.

SN	PSR	Wt Whole Arr	Time	Reg. Analysis

Regression Analysis of Sensor (1)

L-51-I m Osmal = -39.7951 + 53.1371 X μV

9. Measurments of Osmotic Potential

(i) The *Determination of Osmotic Potential of cell-sap by Scholander-Pressure Pump*

Method adopted : Leaves of similar age are collected in aluminium foil and bring to them in laboratory and initial fresh weight is determined and then the leaves are kept in Scholander pressure vessel. The pressure chamber comprises an aluminium, stele or stainless steel pressure vessel that can withstand pressure upto 60 MPa connected to a pressurized supply of inert gas (usually nitrogen or compressed air) through a pressure reducing valve and metaring value and connected to pressure gas.

The top of the chamber can be reported from the cap [K] so that chamber tops quickly interchanged. Different tops may be necessary for grass leaves, leaves with large petioles, leaves with very short petioles or when several samples are measured. The seal [L] itself is important. It is made up of rubber that is sufficient elastic to fill in the indenlatians of irregularly shaped petioles, but not so soft that is disintegrates under pressure. For very irregular petioles, quick setting silicon compound can be used, but this shows down the numbers of leaves that can be measured.

Theory observed : Scholander (1964,1965) described the principles of the pressure^ chamber techniques. They showed that due to the evaporation of water from leaf cells by transpiration, coupled with resistances to flow of water from soil to the leaf, a negative hydrostatic pressure build up in the xylem. When a transpiring leaf is cut, the xylem sap recades in the xylem until it becomes restricted by a cross wall. Scholander et al. (1964) suggested that pressurizing the leaf until water just returns to the cut surface of the xylem tissue gives a measure of the hydrostatic pressure in the xylem.

Scholander et al. (1965) showed that once the cells in the tissue reach zero turgor pressure, Pe is related to the cell water –

$1/Pe = 1 \# S = V_s - V/RTn$

Where, S = is the volume of symplastic water in the turgid leaf.

\# S = is the osmotic pressure of the symplast

V = is the volume of the water lost from the leaf

R = is the Universal gas constant

T = is the absolute temperature and

n = is the number of solute in the symplast

Study of the Pressure - Volumes relations

Another advantage of the pressure chamber techniques is that it can be used to measure the pressurevolume relations to plant tissue. Scholander showed that if additional pressure [Pe] is applied to *a* leaf or twig beyond the initial balancing pressure, xylem sap can be collected and the volume determined. By doing this in stepwise manner, a pressurevolume curve can be established. Once the turgar pressure reached zero, the plot of 1/Pe against V becomes linearas predicted eq. 1. Establishing the pressurevolume curves of species has become popular becomes a wide range tissue water accumulate can be known.

(ii) Determination of Osmotic Potential of cell-sap by VAPRO — VAPOUR PRESSURE OSMOMETER

The VAPRO Osmometer is an advanced electronic adaptation of the hygrometric method of vapour pressure determination. In this instrument, the sensitive thermocouple and sophisticated electronics provide the means to measure the dew-point temperature depression of a specimen with resolution to 0.00031°C.

Vapour pressure and freezing point are among the colligative properties of a solution. Compared with pure solvent, these properties are altered in proportion to the number of solute particles dissolved in each kilogram of solvent (water in the case of biological solutions). Thus, measuring either property is an indirect means of determining solution concentration or osmolality.

The chief advantage of the vapour pressure method is that it does not require alteration of the physical state of the specimen.

Concomitant benefits include:

- 10 microliter sample size.
- Routine operation on microsamples of any biological solution, including whole blood, serum, plasma, urine and sweat, as well as complex specimens such as tissue samples.

- None of the measurement artifacts that arise in freezing point depression measurements due to elevated viscosity, particulate matter, inhomogeneities, or other physical characteristics of the sample.
- Superior reliability because the measurement involves minimal mechanical complexity.

Theory of Operation : Osmolality is an expression of the total concentration of dissolved particles in a solution without regard for particle size, density, configuration, or electrical charge. Indirect means for the measurements of osmolality are afforded by the fact that the addition of solute particles to a solvent changes the free energy of the solvent, i.e., vapor pressure, freezing point, and boiling point. Compared with pure solvent, the vapor pressure and freezing point of a solution are lowered, while its boiling point is elevated, provided that a single solvent is present in the solution. Solutions containing more than one solvent generally behave in more complex ways.

In single-solvent solution, the relative changes in solution properties are linearly related to the number of particles added to the solvent, although not necessarily linearly related to the weight of solute, since solute molecules may dissociate into two or more ionic components. Since these properties all change linearly in proportion to the concentration of solute particles, they are known as "colligative" properties.

Osmotic pressure is also a colligative property of a solution, but unlike the other three, it is not a cardinal property of the solvent. Solution osmotic pressure can be measured directly using a semipermeable membrane apparatus, but only with respect to those solute particles that are impermeable, since smaller solute particles freely transude the membrane and do not directly contribute to osmotic pressure. Such a measurement is referred to as "colloid osmotic pressure" or "oncotic pressure". It is expressed in terms of pressure, in mmHg or kPa. Total osmotic pressure, i.e., that which can be calculated on the basis of total solute concentration, is a theoretical concept only.

The measurement of total solution concentration, or osmolality, can only be made indirectly by comparing one of the solution colligative properties with the corresponding cardinal property of the pure solvent. The first practical laboratory instruments developed for routine measurement of osmolality were based upon depression of the freezing point and, until recent years, all osmometers for large-scale testing were based upon this methodology.

The VAPRO Osmometer embodies newer technology. It is based upon a measurement of vapour pressure depression made possible by thermocouple hygrometry. The vapour pressure method enjoys a significant intrinsic advantage over the measurement of either freezing point depression or boiling point . elevation in that it it can be performed without the necessity for a change in the physical state of the specimen. It is thus a passive technique of measurement that is free from measurement artifacts that often occur when the specimen to be tested must be altered physically. This fundamental difference in methodology gives rise to the many advantages of the vapor pressure bsmometer over the older method.

In the VAPRO Vapour pressure osmometer, a 10 microliter sample of the solution to be tested is pipetted onto a small, solute free paper disc which is then inserted into a sample chamber and sealed. A thermocouple hygrometer is incorporated integrally within the chamber. This sensitive temperature sensor operates on the basis of a unique thermal energy balancing principle to measure the dew point temperature depression within the chamber. This parameter, in itself a colligative property of the solution, is an explicit function of solution vapor pressure.

(a) *Equilibration and zero set:* The sample is introduced into the chamber and the chamber is closed. Simultaneously, "In Process" and a countdown by seconds is displayed. At this point, there will generally be some difference between the temperature of the specimen and the temperature of the sample chamber. Temperature equilibrium occurs within a few seconds. The vapor pressure may also reach equilibrium during this interval. The microvoltmeter reads the amplifier voltage to establish the reference for the measurement.

(b) *Cooling* : An electrical current is passed through the thermocouple, cooling it by means of the Peltier Effect to a temperature below the dew point. Water condenses from the air in the chamber to form microscopic droplets upon the surface of the thermocouple.

(c) *Dew-point convergence* : Electronic circuitry "pumps" thermal energy from the thermocouple via Peltier cooling in such a way as to cancel out heat influx to the thermocouple by conduction, convection, and radiation. Given this, the temperature of the thermocouple is controlled exclusively by the water condensing upon its surface. Thermocouple temperature depressed below the dew point in step @, rises asymptotically toward the dew point as water continues to condense. When the temperature of the thermocouple reaches the dew point, condensation ceases, causing the thermocouple temperature to stabilize.

(d) *End of sequence and readout:* The reading on the display is proportional to the vapour pressure of the solution. When this final reading is reached, a chime sounds and the "In Process" changes to "Osmolality". The result is displayed in SI units of osmolality-mmol/kg.

10. Measurments of Membrane Permeability

It is measured following Fletcher and Drexlure (1980). The per cent leakiness of the leaves are determined by placing 3 leaf discs (1 cm) in 15 ml of distilled water for 24 hours at 24°C. The conductivity of the water is determined using conductance meter. The samples are then placed in boiling water-bath for 20 minutes, cool to 24°C and conductivity is determined a-gain. The membrane permeability is calculated as per cent leakiness as the ratio of conductivity after 24 hours to conductivity after boiling.

11. Membrane Phospholipid

Solutions:

Conc. Perchloric acid : Store in acid bottle away from organic solvents.

ANS solution	:	30 g Sodium Bisulfite NaHSOs 6 g Sodium Sulfite Na2SOs 5 g 1,2,4-Aminopthal Sulfonic Acid (ANS) 250 ml dd water Keep refrigerated in dark bottle. Will keep no more than 3 months.
5% Ammonium Molybdenate	:	5 g Ammonium Molybdenate $(NH_4)_6$ Mo7024 100 ml dd water Keep refrigerated.
ImM Phosphate Std.	:	0.012 g Sodium Phosphate Monobasic NaH-2PO4 100 ml dd water
2:1 Chloroform :Methanol	:	200 ml Chloroform, 100 ml Methanol, Use in fume hoocl.
7% Saline Solution	:	7 g Sodium Chloride NaCl, 100 ml dd water.

Extract Preparation : Grind plant material in 2:1 Chloroform : Methanol. Centrifuge debris at 3000 rpm for 10 minutes. Remove supernatant and place in corex tube. Add 1 ml 7% *NaCl* ('aqueous) for every 5 ml supernatant and vortex. Centrifused at 10,000 rpm for 10 minutes. Remove and store lower supernatant.

Phospholipid Oxidation : Place organic extract containing membranes in graduated 10 ml test tube. Dry under nitrogen to complete dryness. No solvent must be left. Add 0.5 ml of Cone. Perchloric acid and top the tube with a marble. In fume hood, heat in dry block at high temperature for 15 minutes or until solution is clear. Remove from heat and allow to cool for 10 minutes before removing the marbel. Do not touch with naked hand.

Phospholipid Analysis: With the samples, prepare 2 more tubes one marked blank and the other marked standard. Add 0.5 ml Cone, perchloric acid and 1.0 ml of phosphate solution to the standard tube. Add 1 ml of dd water to all of the tubes. Add 0.4 ml ANS solution and 0.5 ml ammonium molybdate to all tubes. Vortex all tubes and incubate at room temperature for 15 minutes. Top all tubes to 10 ml with dd water and vortex. Determine the absorbance at 660 nm blanking on the blank tube. Determine phospholipid concentrations by ratio comparision with the standard :

$$[\text{Phospholipid}]\ \mu M = \frac{Apl}{Astd}$$

12. Study of Porometric

The measurement of stomatal behaviour is fundamental to most areas of plant research. However, these measurements have not always been particularly easy to get, usually involving lessthan-less convenient devices and procedures. Porometer is the instrument which is used to measure the stomatal resistance, transpiration rate, relative humidity, leaf temperature and the quantum received by the intact leaves, instantly. Porometers in-general use today fall into two broad categories, transient or steady state, according to the measurement principle on which they are based. Transient porometers may be ventilate, in which a fan provides rapid chamber mixing, or unventilated, in which cuvette humidity changes occur strictly by diffusion. In ventilated transient porometers, stomatal resistance is calculated from the rate and magnitude of change in cuvette humidity, and system volume. In unventilated transient, porometers, resistance is not computed directly from the magnitude of changes in relative humidity. Four types of ventilated porometers will be considered: these include the flow-through steady state porometer, automatic null balance steady state porometer, and ventilated transient porometer; the fourth type, a recirculating manually controlled null-balance porometer has some of the characteristics of each of the first three types.

Theory/Principle observed : Water loss from a leaf placed in the porometer cuvette is determined by measuring the flow rate of dry air necessary to maintain a constant relative humidity inside the cuvette. Typically, the ambient relative humidity is used as a null point, and dry air is injected into the cuvette at a rate which is just sufficient to balance the transpirational water flux out of the leaf. This maintains the cuvette relative humidity at the set-point. Stomatal resistance (or conductance), is calculated directly from the measured values of relative humidity, leaf and air temperature, and flow rate.

Procedure adopted : The aperture area depending on leaf surface (broad/thin) and atmospheric pressure is initially entered into the porometer. After the cuvette is acclimated to ambient conditions, cuvette relative humidity is stored in memory with ' the help of HUM SET switch, the intact leaf sample is clamped in sensor head. The null adjust valve is used to bring the flow rate within the dynamic range of the automatic flow controller. After thirty seconds, hold switch is used to hold the various parameters in memory for display.

13. Measurments of Chlorophyll Content

100 mg of fresh leaves are homogenised in 80 % acetone and centrifuged for 10 minutes at 3500 rpm. Then the supernatant is made upto 10 ml and the absorbance is read as respected following methods. These are some important methods which are commonly used in research work.

(a) Arnon [1949]

Chlorophyll a (mg/g FW) = 12.7 OD 663 - 2.69 OD 645

Chlorophyll b (mg/g FW) = 22.9 OD 645 - 4.68 OD 663

The values for Chl a and Chl-b are added together to obtain the quantity for total chlorophyll.

(b) Strain and Svec [1966]

Chl a (μg ml^{-1}) = 11.63 (Aees) - 2.39 (A649)

Chl b (μg ml^{-1}) = 20.11 ($A_6$49) - 5.18 (Aees)

Total Chl (μg ml^{-1}) = 6.45 (Aees) + 17.72(Ae49)

14. Measurments of Carotenoid Content

For carotenoid estimation 100 mgof fresh leaf tissue is taken and crush in a pestle and mortar with 80% (v/v) Acetone and a pinch of clean, fine sand. The extract is centrifuged at 2500 g for 10 minutes and tne clear supernatant is made upto 10 ml with 80% Acetone. The absorbance of the extract is read at 440 nm.

Total carotenoids are estimated following the method given by Ikan (1969).

$$Total\ carotenoid\ (mg\ g^{-1}\ F.W.) = \frac{A}{196\ X\ L\ X\ W}$$

Where, A = absorbance at 440 nm

L = diameter of spectrometer quartz cell, and

W = g sample / total volume in ml.

When the absorbance of extract is read at 663, 645 and 480 nms, the carotenoid content can be determined using Kirk and Alien (1965) following the formula –

Car480 = VA480 + 0.114 VA663 - 0.638 VA645

15. Measurments of Stanthocyanin Level

Anthocyanin comprise of a group of glycosides pigments responsible for various colours, particularly red, violet and blue in flower, stem, leaves and roots (beet root) of plant. They are soluble in water and generally occur in the cell-sap.

Anthocyanin is measured using whole plant materials by the method of Arummn and Mohr (1978). Plant material is extracted in 5 ml of HC1: Methanol (1: 99 v/v). The extraction is carried out in darkness at low temperature. The absorbance is measured at 480, 510 and 525 nm in double beam spectrophotometer. Anthocyanin is measured as OD/g material.

16. Measurments of Hill Activity

This can be meaured by using DCPIP Reduction Method.

Reagent required

(I) Composition of suspension medium

(a) 1MSorbitol 10ml
(b) 0.1M K_2HPO_4 7.5 ml
(c) 25mM NaEDTA 6.0 ml
(d) lOmM NaHCOs 6.0ml
(e) Distilled water 0.5 ml (pH should be adjusted at 7.6)

(II) Composition of Assay medium

(a) 1M Sorbitol 15 ml
(b) 75mM Tricine 12ml
(c) 25mM NaEDTA 2.4ml
(d) lOmM NaHCOs 12ml
(e) Distilled water 18.6ml (pH should be adjusted at 7.6)

Isolation of Chloroplast : Take one gram leaf tissues without major veins and grind them in a mortor with a little suspension medium. Make-up the extract to about 30 ml. Filter the extract through four layers of cheese/muslin cloth lined in a funnel. Make up the extract to 30 ml using the suspension medium. Centrifuge at 10.000 rpm at 0°C for 5 minutes. Decant the supernatant and resuspend the chloroplast by adding suspension medium and make up to 6 ml volume and keep it in a freeze. This is an isolated chloroplast suspension.

Assay of Hill-reaction : For estimating the Hill-reaction, dilute the chloroplast suspension using the extraction medium (medium A) so as to get a concentration of 150 ug/ml chlorophyll. Take 2.8 ml of assay medium (medium B) in a cuvvette, add 0.1 ml of diluted chloroplast suspension and 0.1 ml of ImM DCPIP, mixed well by inverting the cuvvette and the absorbance is measured immediately at 600 nm. Next, expose the cuvvette to the light source (using water filter between light source and cuvvette) immediately thereafter for a 5 minutes and measure the absorbance again. Now, by plotting the absorbance values against the standard graph for DCPIP, the amount of DCPIP in the reaction mixture before and after the illumination are be calculated.

The Hill-Reaction is estimated following the further formula –

$$\mu\text{mol DCPIP reduced} = \mu\text{mol DCPIP in reaction mixture} - \mu\text{mol DCPIP remaining after illumination}$$

$$Hill-Activity = \frac{\mu\,mol\ DCPIP\ reduce\ X\ 60}{\underset{in\ minutes}{illumination\ time}\ X\ \underset{rection\ mixture}{\mu\,chl\ in\ the}}$$

17. Measurments of Protein Content

100 mg leaf tissue is crushed in 0.05 N NaOH solution with a pinch of clean fine sand. The homogenate is centrifuged at 3000 rpm for 10 minutes and the supernatant is made upto 10 ml using 0.05 N NaOH.

The protein content in the extract is estimated following Lowry et el. (1951) as modified by McDonald and Chen (1965).

Reagents required

(i) 2% (w/v) Na2COs dissolve in 0.1 N NaOH

(ii) 2.7% (w/v) sodium potassium tartrate

(iii) 1% CuSO4

(iv) IN Folin-ciocalteu's phenol reagent dilute (1:1) with distilled water from a 2N commercial preperation, and make fresh always.

0.6 ml of the extract is taken in a centrifuge tube and dilute with 2.4 ml of distilled water. To this is added 4.5

ml of 10% Trichloroacetic acid (TCA) to participate the protein fraction of the extract. The mixture is centrifuged, the supernatant discards, and the pellet is dissolved in 0.5 ml of IN NaOH and allow to stand for 2 hours for the protein to dissolve. To this is added 5 ml carbonate-copper solution (prepare by adding 0.05 ml 2.7% sodium potassium tartrate and 0.05 ml 1% $CuSO_4$ into 4.9 ml of 2% Na2C03, per test tube. This mixture is prepared in a quantity sufficient for all tubes and while preparing, the CuSO4 solution is always added last. The mixture is always made freshly and incubated for exactly 10 minutes. At the end of 10 minutes 0.5 ml of Folin-phenol reagent is added, mix rapidly and allow to stand for 30 minutes at room temperature. The absorbance of the clear blue solution is noted at 700 hm and the protein content estimate from a standard curve prepare using Bovin Serum Albumin (BSA).

18. Measurments of Free Proline

Free proline is estimated by a simple spectrophotometer method given by Bates et al. (1973).

Reagents required

a.3% (w/v) sulfosalicylic Acid : Sulfosalicylic acid (3%) is prepared by 3 g of sulfosalicylic acid in 97 ml of distilled water.

b.Acid Ninhydrin : The reagent is prepared by mixing 20 ml of 6M orthophosphoric acid with 30 ml of glacial acetic acid. This mixture is heated at 60°C and 1.25 g Ninhydrin is dissolved slowly in this mixture.

c. Toluene.

Procedure adopted : 200 mg fresh leaf tissue is extracted for proline in accordance with Bates et al. (1973) by homogenizing in 10 ml of 3% aqueous sulfosalicylic acid and the homogenate is centrifuged at 5000 rpm for 10 minutes. Two ml of supernatant is reacted with 2 ml of acid ninhydrin. The tubes are incubated at 100°C for 1 hour in waterbath and later transferred into an icebath to terminate the reaction. Mixture extracted with 4 ml of toluene and brought to room temperature. The tubes are agitated vigorously for 15-20 seconds. The absorbance of this Coloured extract is measured at 520 nm using UV Visible Spectrophotometer against a Toluene blank.

The amount of proline in the sample are determined using a standard curve against L-Proline.

19. Measurments of Total Free Amtno Acid

For the extraction and estimation of total free amino acids, the method of Noorudeen and Kulandaivclu (1982) is followed.

(i) Extraction of aminoacids from the leaf tissue

Preweighed material (200 mg) is extracted twice with 80% acetone (total volume 5 ml) after vigorous grinding using a pestle and mortar. The supernatant is used for the estimation of amino acids.

(ii) Estimation of amino acids

Preparation of potassium cyanide-pyridine reagent: 2 ml of 0.01% KCN (10 mg of KGN dissolve in 100 ml of distilled water) and 1 ml of 80% phenol (4g dissolve in 5 ml of distilled water) is added to 100 ml pyridine.

0.5 ml of extract is mixed with 1 ml of KCN-pyridine reagent after which it is kept in a boiling water bath for 10 minutes. To this mixture 0.2 ml of 0.5% ninhydrin in absolute alcohol is added and the tubes are kept in boiling water bath for another 15 minutes. The tubes are cooled to room temperature, the volume is adjusted to 10 ml with 60% ethanol and the absorbance read at 575 nm. Triosene is used as a standard for the estimation of amino acids, and is calculated as follows :

$$C = \frac{K\ X\ A + B\ X\ dilution}{Fresh\ weight\ of\ the\ tissue\ (g)}$$

Where, C = Concentration of total free amino acids in the sample

K = 115.19 B = 1.36

A - Absorbance of the sample at 575 nm

Amino acid content is expressed in mg/g of fresh weight.

20. Measurments of Starch Content

(i) Extraction of starch from leaf tissue

Preweighed material (200 mg) is homogenised with 10 ml of 80% alcohol in a pestle mortar and centrifuged (3500 g) for 15 minutes. The supernatant is discarded and the pelettė is again washed with 10 ml of 80% alcohol and recentrifuged. The supernatant is again discarded and the pellet is dissolved in 10 ml of distilled water and boiled for 40 minute in a water bath to dissolve the starch. The volume is made-up-to 10 ml and the supernatant is collected for starch estimation.

(ii) Estimation of starch

Starch is estimated by the method given by Plummer (1971) with the help of iodine-potassium-iodide (I2KI) reagent.

(iii) Preparation of lzKl reagent

60 mg of Iodine dissolve in 10 ml .of 6% potassium iodide solution. From this 1 ml is taken and added in distilled water along with 0.43 ml of hydrochloric acid (0.05 N) and the total volume is made upto 100 ml with distilled water.

The reaction mixture consist of –

1 ml - Starch extract

4 ml - Distilled water

1 ml - I_2KI reagent

The absorbance of the reaction mixture is measured at 640 nm. The starch content is calculated from calibration curve prepare by using starch as a standard and it is calculated as follows –

$$C = \frac{K\,X\,A + B\,X\,dilution}{Fresh\ weight\ of\ the\ tissue\,(g)}$$

Where, C = Concentration of starch in the sample

K = 4014.7 B= -0.1

A = Absorbance of the sample at 640 nm

Starch content is expressed in mg/g of fresh weight.

21. Estimation of Total Sugar and Reducing Sugar

(i) An extract on of total sugar and reducing sugar from the leaf tissue

The total free sugar (reducing plus non-reducing) is estimated by the method of Fong et al., 1953 (cf : Hassid and Abraham, 1957). Preweighed (200 mg) material is boiled for 10 minutes in 10 ml of 80% (v/v) chilled ethyl alcohol and the supernatant after centrifugation (3250 g, for 15 minutes) is used. The residue is again ground with 10 ml of 80 % (v/v) ethyl alcohol and centrifuge at 3250 g for 15 minutes. The two supernatants are combined and the volume make up to 20 ml. From this alcohol extract, 5 ml is taken for acid hydrolysis with IN HC1 (5 ml) in a boiling water bath for 40 minutes. This hydrolysed extract is used for the estimation of total sugar. From the above alcohol extract, 5 ml is taken and evaporate to dryness, the dry contents are dissolved in distilled water (10 mDand cool in refrigerator for 30 minutes, this extract is used for the estimation of reducing sugars.

(ii) An estimation of total sugar and reducing sugar

A Preparation of Anthrone reagent : 200 mg of anthroiie is dissolved in 100 ml of chilled concentrated sulphuric acid.

The reaction mixture consist of :

2 ml - Total sugar or reducing sugar extract

4 ml - Anthrone reagent

Reaction mixture is kept in boiling water bath for 7.5 minutes to develop colour, the absorbance is taken at 640 nm after cooling. Known concentration of glucose is used as standards and with the help of glucose calibration, total sugar and reducing sugar are calculated as follows :

$$C = \frac{K\ X\ A + B\ X\ dilution}{Fresh\ weight\ of\ the\ tissue\,(g)}$$

Where : C = concentrationof total or reducing sugar in the sample

K= 111.70

B= -2.12

A = Absorbance of the sample at 640 nm.

Sugar content is expressed in mg/g of fresh weight.

(iii) Estimation of total sugar (Malhotra and Sarkar, 1979)

Reagents required

(a) *Copper reagent A* : Dissolve 25 g of sodium carbonate (anhydrous), 25 g of Rochelles salt (SPT), 20 g of sodium bicarbonate and 200 g of sodium sulphate in 800 ml of distilled water and dilute it to one litre. Filter and store.

(b) *Copper reagent B* : 15% of CuSO4.7H2O containing 1 or 2 drops of cone. H2SO4 per 100 ml.

(c) 1N H_2SO4

(d) 1N NaOH

(e) Arseno Molybdate reagent

200 mg of dried leaves are hydrolysed with H2SC-4 at 50°C for 30 minutes. One' ml of the extract is taken and to this one ml of . copper reagent A and copper reagent B are added in the ratio of 25 : 1 ml. Then it is boiled for 30 minutes and one ml of arsenomolybdate reagent is added. Deep blue colour appears ' slowly. The volume is made upto 25 ml and the absorbance is read at 500 nm against blank. The absorbance values are calibered from the standard graph prepare by glucose.

22. Estimation of Non-reducing Sugar

It has estimated by using the method given by Malhotra and Sarkar (1979). For this, 100 mg of dried sample is boiled in

5 ml of IN H2SO4 is added and boil for 5 minutes in a water bath. Then it is neutralised with IN NaOH. 1 ml of extract is taken and the remaining procedure followed is same as that for reducing sugars.

23. Estimation of Total and Free Fatty Acid

Solutions rwquired

(a) 2:1 chloroform : Methanol (v/v)

(b) 7% Salt solution : 7g NaCl + 100ml dd water

(c) 14% Boron Trifluoride in methanol

(d) Haxene

(e) 70 : 30 : 2 Diethyl ether : Petrolium ether : Acetic acid

(f) Chloroform

(g) Heptadecanoic acid : 0.27 g Heptadecanoic Acid (1 μmol/10 Q ml) 2:1 Chloroform : Methanol

Procedure adopted: If whole tissue is to be used homogenize it a ratio of 10: 1 2: 1 Chloroform : Methanol. If free fatty acids are to be measured add 10 ul of Heptadecanoic acid solution. Spin out debris at 3000 g for 10 minutes. Remove supernatant.

If using organelle or microsome pellets, grind the pellet in about 3 ml of 7% NaCl solution. Transfer to corex tube and add 5 ml 2 : 1 chloroform:methanol. If free fatty acids are to be measured add 100 ul of heptadecanoic acid solution.

Centrifuge at 10,000 g for 10 minutes and collect the lower (organic) supernatant.

Total Fatty Acid Analysis: Transfer the organic supernatant to a screwtop vial and dry completely under nitrogen.

Add 1 ml of BFs solution, blanket under nitrogen, vortex and incubate for 90 minutes at 90°C in a dry block heater.

Add 0.5 ml dd water and 1 ml hexane and vortex.

Remove hexane fraction (upper) and place in a 2 ml GC vial

Be careful not to get water in this fraction because it will distroy the GC column.

Dry under nitrogen to about 1/3 volume, blanket under nitrogen and analyze by GLC on a DB17 Megabore column isothermal at 160°C.

Free Fatty Acid Analysis : Transfer the organic fraction to a test tube and dry completely under nitrogen.'

Resuspend the lipids in 5 ml chloroform.

Push the extract through a Sep-Pak Silica column prewetted with 5 ml chloroform.

Ellute the column with 70 : 30 : 2 diethyl ether : Petrolium ether : Acetic acid and pool with chloroform wash.

Transfer the organic supernatant to a screw-top vial and dry completely under nitrogen.

Add 1 ml of BFa solution, blanket under nitrogen, vortex and incubate for 5 minutes at 90°C in a dry block heater.

Add 0.5 ml dd water and 1 ml hexane and vortex.

Remove hexane fraction (upper) and place in a 2 ml GC vial

Be careful not to get water in this fraction because it will distroy the GC column.

Dry under nitrogen to about 1/3 volume, blanket under nitrogen and analyze by GLC on a DB17 Megabore column isothermal at 160° C.

24. Estimation of Ascorbic Acid

The ascorbic acid in the plants is estimated following Schafferts and Kingsley (1955) method.

Reagents required

a. Trichloroacetic acid (TCA) 4% (w/v)

b. 2, 4-Dinitrophenyl hydrazene(2,4-DNPH)-The reagent is prepared by dissolving 2 g 2, 4-DNPH in 100 ml of 9N H2SO4. This mixture is kept for overnight and filter through whatmann No. 42.

C. 85% H2SC-4

d. Thiourea - The reagent is prepared by dissolving 10 g thiourea in 50 ml absolute alcohol. This solution is diluted to 100 ml with distilled water.

e. Norite

Procedure adopted : 500 mg of fresh leaf tissues are extracted in 24.5 ml 4% aqueous trichloroacetic acid and the homogenate is centrifused at 5000 rpm for 10 minutes. 10 ml of supernatant is taken and add 1/4 tea spoon of norite, shake for one minute and filter. 4 ml of the filtrate is reacted with 1 drop of thiourea, 1 ml of 2,4 DNPH, mix well the reaction solution. The tubes are incubated at 80°C for 10 minutes in a waterbath and later transfer to an icebath. The reaction mixture is extracted with 5 ml of 85% H2SO4 mix well and being to room temperature. The absorbance of the coloured extracts is measured at 515 nm against blank. The amounts of ascorbic acid is measured by using a standard curve.

25. Estimation of Tocopherol

The a-Tocopherol in the plants is estimated following Dadlani and Agrawal.

Reagents required

(a) Chloroform : Methanol (2 : 1 v/v)

(b) Ethanol

(c) a a-dipyridyl (0.5 % solution of Ethanol)

(d) Ferric chlorideJ0.2 % solution of Ethanol)

Procedure adopted: (A) Extraction of Lipid : 500 mg fresh leaf tissue is extracted in 10 ml 2:1 Chloroform : Methanol solution and homogenate is centrifuged at 10,000 rpm for 10 minutes. Two layers are seen, lower layer is carefully collect. This is the lipid extract.

Estimation of a-Tocopherol : For drying the extract on medium, it should be carefully evaporated. The best method to achieve, this is to keep the sample in warm water (around 50°C) and blow a slow stream of nitrogen on the surface.

OR

Extract is dried under liquid nitrogen.

Now, add 1.6 ml ethanol, 0.2 ml a,oc-dipyridyl and 0.2 ml of feric chloride into the dried extract. Mix well and keep in dark for 5 minutes, and intense red colour develops in reaction mixture. This reaction mixture is extracted with 4 ml of distilled water. After mixing the solution, two layers are separated out and red[1] colour goes to aqueous layer. The absorbance of the coloured extract is measured at 520 nm against blank.

The amount of a-Tocopherol in the sample is determined using a standard curve generated using a-Tocopherol.

26. Estimation of Nucleic Acid : RNA, DNA

(a.1) Estimation of RNA : RNA estimation is done following the Orcinol reaction (Schneider, 1945). 1 ml aliquot of extract is taken and placed in a test tube, to which is added 1.5 ml of distilled water, 2.5 ml of concentrated HC1 (Containing 0.02% FeCls. 6H20) and 0.15 ml of Orcinol solution (10% in ethanol).

The reaction mixture is placed for 30 minutes in boiling waterbath to develop colour. The mixture is diluted to 7.5 ml or more with distilled water and cooled. Optical density of the solution is read at 660 nm in UV Spectrophotometer against a blank carried through the whole process.

(a.2) Reference solution of RNA : The reference solution is prepared by hydrolyzing commercial yeast RNA (Can be obtained from SISCO, Ranbaxy, Glaxo, etc.) in 0.3N NaOH at 40 mg/100 ml concentration.

The standard curve is prepared for the quantitative calculation of RNA in samples.

(b.2) Estimation of DNA : Dische's diphenylamine reaction (1955) is modified by Burton (1956) is followed for estimation of DNA. Instead of a general concentration of 1.5 g diphenylamine (DNA) per 100 ml of acetic acid, 3g of DNA is dissolved in 100 ml of glacial acetic acid as to obtain greater sensitivity while using this solution, 0.10 ml of aqeuos

acetaldehyde (16 mg/ml) is added for each 20 ml of the reagent.

To the 1 ml aliquot of extract, 0.5 ml distilled water is added. 3 ml of DPA reagent is added to above solution and this mixture is incubated for 18 hours at 25 to 30°C (Kupita et al., 1961). The optical densities of the blue colour, formed on overnight incubation, arc determined at 600 nm and 540 nm against blank, and the difference between the values at two optical densities are used to compute the DNA concentrations. A standard solution is with each test.

(b.2) Reference Solution of DNA : Sodium salt of Calf thymus DNA is used for preparing DNA standard solution. A stock solution is prepared by dissolving DNA at 0.4 ng/ml in 5 mM NaOH. From this, working standards are prepared by mixing measured volume of the stock standard with an equal volume of N-PCA and heating at 70°C for 1.5 minutes. Both standards are stored in the refrigerator. According to Burton (1956) no differentiation of the stock standard occurs even after six months of storage. The working standard is also used in preparing standard curve. With the help of this standard curve the quantitative values of DNA in sample are calculated and multiplied by a factor of 2 or one molecule of DNA is equal to two molecules of deoxypentose sugar.

27. Estimation of Nitrate Reductase Activity

The nitrate Reductase Activity is measured following the method of Srivastava and Mathur (1980).

Preparation of reagents required

a. 0.02 % NED N-(l-naphtyl ethylene diamine dihydro chloride) prepare in distilled water (w/v)

b. 1 % sulphanilamide (w/v) prepare in 1 N HC1

c. 0.1 % KNOa solution prepare in distilled water -

d. 5% n-propanol

e. Phosphate buffer (pH 7.4) - It is prepared by mixing 19 ml of 0.2 M KH_2PO_4, the mixture dilutes to 200 ml.

f. Assay Mixture :

The assay mixture is prepared as –

Phosphate buffer	- 4.0 ml
0.1 M KNO3	- 0.5 ml
5% n- Propanol	- 0.5 ml
Total volume obtained	- 5.0 ml

Procedure adopted : 200 mg of leaves of each sample are sliced into 2 to 3 mm fragments and placed in a light proof serum vial, in which 5 ml of assay mixture is already placed. The vials are incubated in dark at 30°C for 30 minutes.

The nitrate formed is estimated by adding 1 ml of 1% sulphanilamide and 1 ml of 0.02 % NED to 1 ml of assay mixture after incubation. The absorbance of the pink colour developed, is measured at 540 nm in spectrophotometer. The enzyme activity is estimated in terms of nitrite (um N02) produced per gram fresh weight of leaf tissue per hour.

$$\text{Enzyme activity} = \mu m\ NO_2\ g^{-1}\ FW\ hr^{-1}$$

A concentration of nitrate formed is calculated by calibration curve prepare by using sodium nitrate solution as a standard.

28. Estimation of Assay of Nitrogenase in Root Nodules of Leguminous Plants

The nitrogenase (EC 1.19.2.1) activity in root nodules can be determined following the method of Srivastava et al. (1980). For this, root nodules are detached from the roots and 100 mg nodules are placed in a light proof 15 ml capacity serum-vial in which 0.75 ml of 100 mM Tris-HCl buffer pH 8.0 and 0.05 ml Toluene are added. The vial is then sealed with a vaccine stopper by using rubber solution. The stopper contained an abraded glass rod (2 cm. long x 4 mm diameter) freely suspended in its middle and the tip is acidified by dipping it into 1 N H2SO4. The excess acid is removed with a filter paper.

The vial is evacuated and then freeze in a deep freeze *(0°C)* for 3 h. After thawing the content at room temperature, 0.75 ml assay mixture, which contains in a final concentration of 13.3 mM MgC12, 3.3 mM ATP and 20 mM sodium dithionite are injected. The reaction is initiated immediately by injecting 8 ml of nitrogen gas in the vial and incubation is done for 10 minutes at 30°C. The reaction is stopped by injecting 1 ml of saturated K^COa solution. The vials are shaken (60 strokes per minute) in an automatic shaker for 8 h for completing release of ammonia. The ammonia trappes at the tip of glass rod in the form of ammonium sulphate is determined colorimetrically with Nessler's reagent at 440 nm using spectrophotometer. The value of ammonia is calculated with standard curve drawing from ammonium sulphate.

29. Estimation of Peroxidase Activity

(i)This can be measured following Shanon et al., (1966)

Reagents required

a. Acetate buffer pH 5.5 (0.1M)
b. Orthodianiside 0.5 % in ethyl alcohol
c. 3% H_2O_2
d. 5 N H_2SO_4
e. 0.9% KC1

50-100 mg of fresh leaf samples are extracted for enzyme in 5 ml 0.9% KC1. Then it is centrifused at 15,000 g for 15 minutes. All the estimations are carried out in ice baths. To 50-100 ul of enzyme extract 2.5 ml of acetate buffer, 0.1 ml diansidine, 0.1 ml H2O2 are added and leave for 30 minutes. The reaction is then terminated with 0.5 ml H2SO4 (5N). The absorbance is read at 400 nm against blank. One enzyme unit is defined as change in 1 OD per minute under the assay conditions.

(ii) : It can also determine by using another method given by Maehley (1954) and George (1953).

Procedure adopted : 200 mg of the material is taken and crush with sterilized sand in a pre-chilled pestle and mortar with 5 ml of chilled acetone in cold room at 0°C and centrifuge at 5000 g for 20 minutes. The supernatant is discarded and the pellet is redissolved in 5 ml of chilled Na-phosphate buffer (0.02 M, pH 6.4) keep for 15 minutes with occasional shaking and again centrifuge for 20 minutes. The supernatant is used for the extraction of wall bound enzyme.

Process of Determination of cytoplasmic peroxidase activity

This is measured by the method of Maehley (1954) and George (1953), using 20 mM guaiacol as an electron donor in presence of substrate H2O2- The reaction mixture is prepared as under—

1ml	- enzyme extract
3ml	- sodium phosphate buffer (0.02 M, pH 6.4)
1ml	- Guaiacol (20 mM)
0.03 ml	- H2O2 (1 % v/v)

The changes in absorbance is recorded at 470 nm. The activity is expressed as AA 470 nm min" mg" protein.

Method of extraction: of wall bound peroxidase : The pellet is washed four times with sodium phosphate buffer or till the washing is free of cytoplasmic peroxidase. Then this pellet is resuspended in 5 ml of 1 M NaCl solution (Ridge and Osborne, 1970) for 1 hour and centrifuge for 15 minutes at 5000 rpm. The supernatant serves as a source of wall bound peroxidase.

Process of determination of wall bound peroxidase activity : The activity is estimated as described in cytoplasmic peroxidase following the method of Maechley (1954) and George (1953). The activity is expressed as AA 470 nm min^{-1} mg^{-1} protein.

30. Estimaton of Superoxide Dismutase Activity

This can be measured by using the method given by Keith et al.,1983.

Reagents required

a.50 mM phophate buffer pH 7.0 containing 1% polyvinyl pyrrolidine.

b.0.2 M potassium phosphate buffer pH 8.6 containing 10 EDTA and 2 x 10 M cytochrome C (substrate).

c. Alkaline dimethylsulfoidc (DMSO containing 5 mM NaOH).

Method observed : 200 mg of the leaf tissue is extracted in 50 mM phosphate buffer (3 ml) and is centrifuged at 30,000 g for 30 minutes. To the 0.2 ml of the extract one ml of the substrate is added and keep for 30 minutes. Then the reaction with 0.5 ml alkaline DMSO. The blank is terminated with DMSO containing no NaOH. The absorbance of the reduced cytochrome is measured at 550 nm against samples prepare under similar conditions except that DMSO do not contain NaOH. A unit of enzymatic activity is the amount of enzyme which causes 50% inhibition of alkaline DMSO mediated cytochrome C reduction.

31. Estimation of Sulphate Content

0.2 g Dried ground leaf sample is taken in a 50 ml Kjeldahl flask and is digested with a mixture of 0.1 g selenium oxide, 10 ml of cone. HNOs and one ml of cone. HC1. To the residue 0.2 ml of dilute HC1 is added and again evaporated to dryness. One ml cone. HC1 and 25 ml distilled water is added and then boil and filter in 100 ml volumetric flask. The filtrate is made to 50 ml by 5 ml 3 % glycerol and distilled water.

Aliquiot of 10 ml is taken in a test tube and 0.5 ml of 2% Barium chloride solution is added. Sulphate is precipitated as Barium sulphate and the absorbance is read at 420 nm against blank. The actual sulphate content is calibrated from the standard curve prepare by potassium sulphate solution and is expressed as mg/g dry wt.

32. Estimation of Total Nitrogen

An estimation by using the method given by Me Donald (1978). 10 g of oven dried plants material is digested in corning test tube with 12 drops of con. H2S04 and a pinch of digestion mixture, having with the under mastered chemical.

2.0 g	Potassium
1.5 g	Sodium thio sulphate
1.5 g	Salicylic acid
1.0 g	Selinium.

Sometimes cooling the digestion mixture add a few drops of hydrogen peroxide and heat for some time. Finally the colourless digested fluid is made up by distilled water to 10 ml volume. Then to this 0.2 ml of sodium silicate (10%) and sodiumhydroxide (10%) are added and shake thoroughly. The final volume is made to 10 ml with distilled water. The 2 ml of diluted (1:2) Nessler's reagent is mixed and shake. Then after 10 minutes the absorbance of yellow is read at 425 nm. The amount of nitrogen is calculated with the help of standard curve prepare from ammonium sulphate. The results are expressed as percentage.

33. Estimation of Sulphurdioxide Generation

A gaseous pollutant, sulphurdioxide is generated by bubbling air through 1% aqueous sodium metabisulphite solution at a fixed rate. The air flow rate is first standardized in a 1 x 1 x 1 m polythene open top chamber. The generated gas is then measured at one hour interval with Toxic gas monitor. The avearage value obtained is considered as the concentration of the gas to which the plants are exposed. The amount of the gas generated is dependent on the air flow rate.

34. Estimation of Tdf (Total Dust Fall)

Total dust fall is estimated with the help of uniform size jars. Five jars are placed at each site at an height of 8 meters from ground surface, open to sky for two months. They are then brought to the laboratory and settled dust is weighed and express as total dust fall g/m /day.

35. Estimation of Leaf Wash Ph and Leaf Extract pH

Leaves are washed in 50 ml of distilled water twice and the pH of the water is determined by Digital pH meter directly. For estimating the Leaf extract pH, homogenise 5g fresh leaves in a mortar with 25 ml of double distilled water and measure the pH of homogenite directly by Digital pH meter.

36. Estimation of Tgm Study (Toxic Gas Monitor)

Principle observed : A Toxic Gas Monitor measures the concentration of a particular gas in the air by monitoring the amount of colour change produced in the analyzers reagent systems. Air to be analysed is continuously drawn into the instrument via an internal vaccum pump and is made to pass through proper absorbing reagent via series of tubings and the colorimetric estimation of the gas is displayed on the digital read

Procedure adopted : *(i) Sulphurdioxide* : Suphur dioxide estimation in the embient air is done by continuously absorbing the air at an fixed rate. The air sucked by the pump is then reacted with pararosaniline and formaldehyde reagent to form intensly coloured pararosaniline methyl sulphonic acid, whose intensity is measured at 550 nm. The intensity of the colour produced is proportional to the concentration of S02 absorbed.

(ii) Nitrogendioxide : An air is absorbed into the azodye forming reagent with chromium trioxide to convert NO present to NO2- The intensity of the azodye formed is measured at 550 nm.

A digital read out for both the gaseous pollutants are calibrated from the standard graph supplied with the monitor.

37. Estimation of Hvs Study (High Volume Sampler)

Principle observed : Air at a predetermined rate is sucked in by a cyclone (motor) through a filter and there by making the suspended particulates in ambient air adsorb to the filter. With initial and final weight of the filter paper the amount of suspended particulate is estimated in the total volume of the air sucked in.

Procedure adopted : Air at 0.5 LPM rate is sucked in by the motor through a 250 x200 mm, 0.3 micron glass fibre filters. The initial airflow rate and weight of the filter paper is noted. The sampler is run for 8 hours at each site. Final air flow rale and weight of filter paper is noted. Mass concentration of SPM is evaluated by the formula

$$SPM = \frac{[Wf - Wi]}{V} X 10$$

Where, Wi and Wf are the initial and final weight of the filter paper (g) and V is the air volume in cubic meters.

38. Study Cps (Cascade Particle Separator)

Principle observed : The principle remains same as that of the high volume sampler, only the different sized suspended particulates flow from ring nozzle of stage and are separated by crushing and attacking on the surface by impactor method.

Procedure adopted : Filter papers with differential retention capacity are placed separately on different ring nozzle of stage. Air is sucked in at the flow rate of LPM through these filter papers. The sampling is done for 8-24 hours at various sites. The various particle fractions that can be measured'are, <0.7,0.7-1.6, 1.6-5.4, 5.4-10.9 and >10.9 μ.

39. Study Aas (Atomic Absorption Spectrophotometer)

Principle observed: Light at the resonance wavelength of initial intensity I_0 is foccused on the flame cell containing ground state atoms. The initial light intensity is decreased by an amount determined by the atom concentration in the flame cell. The light is then directed on to the detector where the reduced intensity "I" is measured. The amount of light absorbed is determined by comparing I to I_0.

Hollow cathode lamps are used as light source as they are designed to emit the atomic spectrum of a particular element, and the specific lamps are selected for use depending on the element to be determined.

The specific limitations and conditions for each element are as follows :

Element	Wavelength (nm)	Slit width (nm)	Flame Gases	Sensitivity check @
Cadmium	228.8	0.7	A-Ac	1.5
Chromium	357.9	0.7	A-Ac	4.0
Cobalt	240.7	0.2	A-Ac	7.0
Copper	324.8	0.7.	A-Ac	4.0
Iron	248.3	0.2	A-Ac	5.0
Lead	283.3	0.7	A-Ac	20.0
Manganese	279.5	0.2	A-Ac	2.5
Nickel	232.0	0.2	A-Ac	7.0

A - Ac = Air - Acetylene gas; @ = Metal concentration (rhg/1) in aqueous solution which will give a reading of approximately 0.2 absorbance units.

40. Tissue Culture Methodology Technology "In Vitro Culture"

In vitro culture, in which isolated organs, tissues, cells and even protoplasts (cells without wall) of plants are grown in nutrient medium in laboratory under aseptic conditions, is one of the growth areas of biotechnology.

Types of In vitro Culture of Plants

(i) Culture of (almost) whole plants

(i) *Seed Culture :* A seed may be sown in vitro from which a seedling, and finally, a plant developed, e.g. Orchids.

(ii) *Embryo Culture* : An excised embryo is grown.

(ii) Organ Culture

An isolated organ is grown. Different types can be distinguished, e.g. anther culture, ovary culture, nucellus culture.

(iii) Tissue Culture

A differential tissue is isolated and grown.

(iv) Single cell Culture

The growing of individual cells which have been isolated from a tissue, callus or suspension culture.

(v) Protoplast Culture

The culture of protoplasts (cells without outer wall).

Apart from the actual subject matter of "in vitro culture" technology of (almost) whole plant, organ, tissue, cell and protoplast culture. There are three very important aspects which deserve mention first. These aspects are *nutrient media* maintenance *of aseptic condotions* and *aeration.*

Preparation Nutrient Media

Nutritional requirements for optimal growth of plant organ, tissue, cell and protoplast *in vitro* generally vary from species to species. Even tissues from different parts of a plant may have different requirements for satisfactory growth. It is, therefore, no single medium as such can be suggested as being entirely satisfactory for all types of in *vitro* cultures. When starting with a new system, it looks essential to work out a medium that would fulfil the specific requirements of that system. This is why many recipes of nutrient media have been developed (Table 1).

In the earlier time in vitro culture media, e.g. root culture medium of White (1943) and callus culture medium of Gautheret (1939) were developed from nutrient solutions earlier used for whole plant cultuere. White evolved the medium from Uspenski and Uspenskaia's medium (1925) for algae, and Gautheret's medium is based on Knop's (1865) salt soluton. All subsequent media for mutations are based on White's and Gautheret's media.

A suitable medium formation for a new system, it would be better to proceed with well-known basal medium such as MS (Murashige-Skoog) or Bs (Gamborg et al.). By making minor qualitative and quantitative changes, through a series of experiments, a new medium may be evolved to accommodate specific requirements of the plant materials in question.

PREPARATION OF MEDIA

A. *Stock* - A : In a conical flask, 200 ml of distilled water is taken. All the macronutrients are weighed and are added serially till it completely dissolves. Similarly, in another conical flask containing 200 ml distilled water, all the micronutrients are weighed and are dissolved serially. Finally make up volume to 500 ml.

B. *Stock - B* : Similarly, all organic compound (vitamins, auxins, cytokinins) are added as desired. About 3% sucrose is added. Finally the volume is made upto 500 ml.

C. Add both the stock (i.e. Stock A & B) and a clear and transparent solution is obtained. pH is adjusted using NaOH, HC1 or KOH.

D. Finally agar is added of about 0.8% and is heated until the clearity of the solution is obtained.

E. The prepared culture medium is poured in a culture tubes.

F.Mediu m is sterilized in a validated autoclave at 1 kg/cm (15 psi) 121°C, for the time period. according to the volume.

G. Medium is allowed to cool and it gets solidify (if an agar medium).

Table XI 1. Media Amount (in mg/1)

Component	White (1943)	Heller (1965)	Murashige & Skoog (1962)	Gamberg *et al.* (1969)	Nagata & Tkebe (1971)	Nitsch (1969)
(A) Macronutrient						
KCl	65	750	-	-	-	-
$(NH_4)_2SO_4$	-	-	-	134	-	-
NH_4NO_3	-	-	1650	-	825	720
$NaNO_3$	-	680	-	-	-	-
KNO_3	80	-	1900	2500	950	950
$Ca(NO_3)_2$	300	-	-	-	-	-
$CaCl_2.2H_2O$	-	75	440	150	220	-
$MgSO_4.7H_2O$	720	250	370	250	1233	185
$Na_2SO_4.10H_2O$	200	-	-	-	-	-
KH_2PO_4	-	125	170	-	680	68
$NaH_2PO_4.7H_2O$	16.5	150	-	150	1233	-
$CaCl_2$	-	-	-	-	-	166
(B) Iron						
Na_2EDTA	-	-	37.3	37.3	37.3	37.3
$FeSI_4.7H_2O$	-	-	27.8	27.8	27.3	27.8
$FeCl_3.6H_2O$	-	1.0	-	-	-	-
$Fe_2(SO_4)_3$	2.5	-	-	-	-	-

Contd...

Table XI.1 contd..

(C) Micronutrients						
$MnSO_4.4H_2O$	7	0.01	22.3	-	22.3	2.5
$MnSO_4.H_2O$	-	-	-	10	-	-
$ZnSO_4.7H_2O$	3	1	8.6	2	8.6	10
H_3B0_3	1.5	1	6.2	3	6.2	10
KI	0.75	0.01	0.83	0.75	0.83	-
$a_2MoO_4.2H_20$	-	-	0.25	0.25	0.25	0.25
$CuSO_4.5H_2O$	-	0.03	0.025	0.025	0.025	0.025
$CoCl_2.6H_2O$	-	-	0.025	0.025	-	-
$CoS0_4.7H_2O$	-	-	-	-	0.03	-
$NiCl_2.6H_2O$	-	0.03	-	-	-	-
$AlcL_3$	-	0.03	-	-	-	-
(D) Organic Compound						
Biotin	-	-	-	-	-	0.05
Meso-inositol	100	100	100	100	100	100
Nicotinic acid	0.5	-	0.5	1.0	-	5
Pyridonine-HCl	0.5	-	0.5	1.0	-	0.5
Thiamine-HCl	0.1	1.0	1.0	10.0	1.0	0.5
Glycine	30	-	2.0	-	-	2.0
Ca-D-pantothenate	1.0	-	-	-	-	-
Folic acid	-	-	-	-	-	0.5

Contd...

Table XI.1 contd..

(E) Carbohydrate						
Sucrose	2&	2%	3%	2%	1%	2%
Mannitol	-	-	-	-	13%w/v	-
(F) Growth Regulator						
Kinetin	-	0.25	0.04-10	1.0	-	-
2,4-D	6.0	1.0	-	-	-	-
IAA	-	1-30	-	-		
NAA	-	-	-	-	3.0	-
6-Benzyl-aminopu rine	-	-	-	-	1.0	-
pH	5.5	-	5.7-5.8	5.5	5.8	-

Volume (ml)	Minimum Sterilization time at 121°C
20-50	15 min.
75	20 min.
250-500	25 min.
1000	30 min.
2000	40 min.

Stock solutions af Preparation Plant Growth Regulators

A. Stock solutions of plant growth regulators are prepared by first dissolving them in a few drops of IN NaOH or IN HC1 and then making up the volume with distilled water.

B. mg/ml stock solutions of plant growth regulators is prepared by dissolving 50 mg in 50 ml of distilled water and is stored in refrigerator.

C. For using micromole concentrations, it is convenient to make a 10 M solution of which 1 ml is added to the 1 1 of medium and will give 1 (iM concentration of the compound.

For e.g. to make 10 M stock of IAA (M.W. 175) 17.5 mg is weighed and is dissolved in a volumetric flask of 100 ml, in a few drops of IN NaOH or ethanol. Finally the volume is made up with distilled water and is stored in refrigerator.

D. Among plant growth regulators, IAA is protected from light by covering the bottle with black paper as it degrades in light.

E. All stock solutions must be used within four weeks of preparation.

F.Sol ution should be discarded if it gets precipitated or there is any colour.

ASEPTIC CONDITIONS

The nutrient media used in *in vitro* culture studies, specially when they contain sugar, attract variety of microorganisms and help them luxuriently growing in it. The microorganisms grow at the faster rate than the cultured tissue in medium and finally, kill the latter. It is, therefore, obligatory that a completely aseptic condition is maintained in and around the culture equipments so that the microorganisms fail to contaminate the nutrient media. Following of the three main sources of contamination of the media, and the subsequent methods to overcome them are -

A. The microorganisms may be present in the nutrient medium right from the beginning. These microorganisms are destroyed by properly plugging and autoclaving (steam heating under pressure). The medium can be completely sterilized by maintaining it at 121°C for the time period according to volume.

B. Finally, strict precautions are taken to prevent the entry of microorganisms when the plug of a culture is removed to transfer the plant material to the nutrient medium or from one medium to the other. For this, all operations are done in completely aseptic conditions.

SURFACE STERILIZATION

The plant part to be cultured may carry microorganisms with it. To avoid it, the plant part is surface sterilized as bacterial and fungal growth are detrimental to culture. Explants are washed in mild detergent before treatment with the disinfactory solution, they are rinsed thoroughly under running tap water for 15-30 min. and is submerged into the disinfectant solution and is agitated gently. Disinfectant solution commonly used are (a) Calcium hypochloride (9-10%) for 15-30 min. and (b) Mercury chloride (0.1-1%) for 5-10 min.

Sterilization is enhanced by placing the material in a 70% ethyl alcohol solution prior to treatment with another disinfectant solution. The use of two steps (two sources) sterilization procedure is of more beneficial. A wetting agent such as Tween-20 or 80 is used to reduce surface tension and allow better surface content. It is added along with the disinfectants under sterile conditions, the solution is decant and explants are rinsed with sterile distilled water several times. If it is conducted under vaccum, this results in the removal of air bubbles and provides a more efficient sterilization proccess.

INNOCULATION

Air flow system of laminar is switched on for about 1 hr and UV lamp for about 15 min. before innoculation. Hand should be washed with soap and water. The surface of the laminar bench should be wiped with a swab of spirit. All the requirements such as spirit lamp, forceps/scapel in a beaker half filled with spirit, surface sterilized material, sterile petridishes, sterile distilled water and sterile medium, all are arranged on a bench according to thc convenience in working. The forceps and scalpel blades are flamed and are kept for cooling on a cooling rack or sterile petridish. Plant material is then cut into pieces of desired size in a sterile petridish. The tube or flask is opened near the flame and mouth of the tube is heated for sometime so that viral or bacterial contamination can be avoided. The cut explaht are innoculated on the medium with the help of forcep and is pressed lightly and plug is replaced. All the steps are carried out quickly to avoid the contamination.

AERATION METHODOLOGY

While *in vitro* culturing is going on the cultured plant part should be given proper aeration. If the plant part is grown on the surface of a semi-solid medium it acquires sufficient aeration without a special device. However, if liquid medium is used the plant part, since it gets submerged, requires some special device for proper aeration. One device is to use a "filter paper bridge" whose two legs remain dippingin the medium, and the horizontal part carrying the plant part is raised above the level of the medium. Agetation of the medium by passing through it filter sterilized air also serves the purpose. More commonly, however, aeration is provided by shaking the flask or tubes oh an automatic shakes. Besides supplying 0 to the plant part, shaking of the medium also brings about dissociation.

of the tissue into small clumps and even single cells. The importance of single cells in the cloning and higher plant genetics has-been, greatly appreciated.

41. Measurement of Metal Tolerance and Phytotoxicity

Every organisms, (Plants or Animals), regardless of whether it lives in a metal-enriched environment or not, has certain ability to cope with non-essential metals or excessively available essential metals, although, there are limits in the occurrence of metals tolerance in a lot of higher plants. Several workers have referred to this ability as "tolerance" or "resistance". It is to restrict these terms to cases of a heritable increase in the ability to cope with excessive metal, usually naturally or artificially selected under the pressure of a toxic level of metal exposure. Such heritable tolerances occur in a limited number _of_ plant species and have been demonstrated for Zn, Pb, Cu, As, Cd and other metals. -Evolution of tolerant population may occur as rapidly as within 5 to 10 years, or more slowly dependent on the selection pressure prevailing.

Tolerance to metals can be demonstrated at different levels of integration. Basically metal tolerance is a constitutive property, present in every living cell, tissue and organ of the plant and animals. It is maintained in cell tissue cultures prepared from tolerant plants. Therefore, tolerance can be tested using different tissue, organs or whole organisms in case of plant. Although, it can not be excluded that process taking place at integration levels higher than the cell may substantially contribute to the tolerance of the whole plant.

There are four methods for the measurement of tolerance and phyto-toxicity in plant systems :

A. Comparative plasmology
B. Rooting technique
C. Whole plant testing
D. Chlorophyll synthesis

Method-1: Techniquie used in Comparative plasmology

First time Repp (1983) elaborated a test for plasma membrane integrity after exposure to toxic metals. Sections of leaf or stem tissue, between 2 to 4 cell layer thick, are incubated are series of metal concentrations, either nutrient addition or a nutrient solution. After 5 to 6 hours (in some cases 24 hours) the cells ability to plasmolyse or deplasmolyse is tested using 1 M glucose.

OR

After keeping the test materials in metal solution nearly 5 to 6 hours solute leakage will be measured by any sensitive conductimeter. The solute leakage will. expressed as the difference between the initial and final conductivity of the solution.

The advantage of this technique is that it allows testing of established plants in field and in the laboratory without cloning them.

Method -2 : Technique used in Rooting

Roots are more directly confronted with toxic metals in the environment than shoots, except in the case of aerial metal deposition. Therefore, root growth usually responds more rapidly to metal exposure than shoot growth. This is probably the reason that root growth is the most widely used parameter in metal tolerance tests. A commonly used measure of tolerance is the tolerance index.

$$\text{Tolerance Index} = \frac{\text{Root growth in the metal solution}}{\text{Root growth in control solution}}$$

The use of tolerance index certainly allows a perception of the coarse patterns of interspecific and intraspecific variation in metal tolerance. The resolution of this index, however is limited and depends on the metal concentration chosen for testing, as well as the range of variation in tolerance present among the plants to be tested.

Moreover, low metal concentrations may stimulate root growth, specially in case of tolerant plant.

Method 3 :Technique used in Whole plant testing

Several workers have used the relative growth rate of the whole plant as an effect parameter in metal tolerance tests. In view of the fact that tolerant and non-tolerant plants differ may differ in the rate of translocation of metals in the roots. It is conceivable that they may also differ in the degree of shoot growth inhibition, relative to the degree of root growth inhibition, if they exposed to toxic metal concentration. If so, then whole plant testing may be expected to differ in many attributes from those obtained_with a rooting test.

Method 4 : Technique used in Chlorophyll synthesis

This is one of the best method for the measurement of the toxicity and tolerance of plant in relation to heavy metals. Photosynthesis is an important process of plant in which light energy is converted into chemical energy. The chlorophylls have long been recognized as the primary light acceptors in plants and they are invariably present in every organism which carries out photosynthesis with absorption of carbondioxide and evolution of molecular oxygen. Heavy metals inhibit the photosynthesis in various ways. Stunting and chlorosis are commonly observed reactions of metal toxicity. One of the commonly reported symptoms of Cadmium toxicity is the chlorosis of leaves. It has been proved that any concentrations of toxic metals showed phytotoxicily to the plants.

12

Technique Used in Water Analysis

For Temperature Measurments

(i) *General aspect:* Temperature plays an important role in biological activities, fish life, limnological studies etc. Temperature governs solubility of oxygen, carbon dioxide, bicarbonate carbonate equilibrium.

(ii) *Apparatus required :* A thermometer having a quick response and having atleast divisions for 0.5°C.

(iii) *Procedure adopted :*

(i) *Air temperature* (ambient temperature) Hold a dry thermometer at least one meter above ground, keeping it away from any heat source or direct solar radiation and read.

(ii) *Water Temperature :* Thermometer is dipped directly into water body and a constant reading is noted. Temperature can be measured in the sampling bottle immediately after taking the sample. Ensure that container attains the temperature of the water sample.

(iv) *Expressions of results :* Temperature may be recorded to the nearest of 0.5′C.

(v) *Precautions observed :* While taking temperature of water sample ensure that the mercuiy bulb of the thermometer is dipped in the water. Before making any temperature measurement, wipe the thermometer with tissue paper or filter paper. Whenever the embient air and water temperature both are to be measured, it is suggested that the earlier should be followed by the later.

Calculation of Ph Value

(a) *General* : pH is the negative logarithm of hydrogen ion concentration (activity of hydrogen ion) in moles per litre at a given temperature. The pH of natural water is controlled by the carbon dioxide/bicarbonate equilibria and usually ranges from 1.0 to 9.0. The majority of water are slightly basic due to the presence of bicarbonates and carbonates. pH affects the chemical and biological properties of liquids and hence is an important estimation; e.g. for controlling corrosion, and water and waste treatment processes .

Electrometric and colorimetric methods are available for the measurement of pH.

Electrometric technique

(a) Principle observed :

The pH is determined by measurement of the electromotive force of a cell comprising an indicator electrode, an electrode responsive to hydrogen ions (glass electrode) immersed in the test solution, and a reference electrode (usually colomelelectrode). Contact between the test solution and the reference electrode is achieved by means of a liquid junction, in the reference electrode. A difference of 1 pH unit produces a potential charge of 58.16 mV at 25°C. The electromotive force is measured with a pH to read directly as pH value.

Interferences : The glass electrode is free from interferences such as colour, turbidity, collodial matter, oxidant reductants. In case, salinity is high a sodium ion error at pH,10 and above has to be compensated. Temperature affects the potential at electrodes and the ionization in the sample. The first effect can be overcome by a temperature compensation adjustment provided on the most of the instruments. The second effect is inherent inthe sample and is taken in to consideration by recording both temperature and pH of each sample.

***(b) Apparatus required*:**

(i) *pH Meter* - Mains or battery operated model (for field measurement) with temperature compensation adjustment and reading to an accuracy of 0.1 pH units.

(ii) *Glass electrode* – This must be compatiable with the pH meter used and should cover the entire pH range with minimum sodium ion error.

(iii) *Reference electrode* - Calomel electrode is widely used. Less concentrated solutions of potassium chloride (e.g. 3.5 M KCl or 350 g/1) are more satisfactory than the saturated solution usually used which very often leads in clogging of the £, electrode.

***(c) Reagents required*:**

Standard pH buffer solutions : This can be prepared from buffer tablets which are commercially available. However, an analyst can prepare his own solutions by referring to standard text book.

***(d) Procedure adopted*:**

(i) Rinse the electrodes with distilled water and dry by gentle wiping with a soft tissue paper.

(ii) Standardize the instrument with the electrode immersed in a buffer solution with a pH close to that of the water to be tested.

(iii) If manual temperature compensation is provided, set it to the temperature of the buffer.

(iv) Check the electrode response occasionally by measuring a second buffer solution of different pH.

(v) Rise with distilled water and dry with tissue paper after every measurement. The difference in reading for the standard buffers should not exceed 0.1.

(vi) Immerse the electrode in the sample.

(vii) Set the temperature control adjustment knob to the temperature of the sample. Taking final reading after 30 seconds.

(e) Expressions of Results :

pH value should be reported to the nearest 0.1 pH unit. The temperature of the sample at the time of the pH measurement should be indicated.

(f) Precision and Accuracy :

The precision and accuracy attainable with a pH meter depends on the type and condition of the instrument. A precision of ± 0.02 pH unit and an accuracy of ± 0.05 pH unit can be achieved. However, ±0.1 pH is satisfactory. Hence, pH value should be reported to the nearest of 0.1 pH unit.

(g) Precoutions taken :

(i) Before making any measurement rinse the electrodes with distilled water and dry it with the tissue paper.

(ii) Ensure adequate level of filling solution in the reference electrode.

(iii) Always keep the electrode in distilled water, never allow them to dry out.

(iv) Standarize the instrument in case of any doubts with the pH buffer closer to the pH of the test sample. .

(v) Adjust the temperature compensation knob before reading.

(vi) After immersing the electrode, wait for sometime so as to stablize the system. Take the reading as soon as it becomes constant.

(vii) As for as possible it should be measured at the site or as soon as possible.

(viii) No chemical preservation of sample is permissible.

(ix) Universal glass electrodes ensure good results.

3. MEASURMENTS OF ELECTRICAL CONDUCTIVITY

(a) General Method :

Conductivity is a measure of the ability of a water sample to £arry an electrical current and it depends on the ionic strength of the water. This mostly depends upon the nature of the various dissolved ionized substance, their actual and relative concentrations and the temperature at which the measurement is made.

Most of the inorganic acids, bases and salts (such as hydrochloric acid, sodium carbonate and sodium chloride) are relatively good conductors. Conversely, most organic compounds, such as sucrose and benezene that do not dissociate in aqouous solution conduct a current very poorly, if at all.

The determination of electrical conductivity is a rapid and convenient means of estimating the concentration of electrolytes. In waters containing mostly mineral salts (waters used for public supply and many other ground or surface waters) this will not be very different from that of the dissolved solids. Hence, in such a samples conductivity can be used as a measure of dissolved solids.

For a particular samples one can have a factor to convert electrical conductivity to dissolved solids. This factor will very from sample to sample. Thus, in some cases measuring the conductivity and converting it into dissolved solids, can be used as a check for the quality of the effluent discharged or for monitoring the treatment plant. The factor usually varies from 0.55 to 0.9. The factor is multiplied by conductivity (umhos/cm) to obtain dissolved solids. Some other important, applications of conductivity measurements are :

(i) The purity of distilled and deionized water can be checked by conductivity determination.

(ii) To check the variations in the dissolved mineral concentration of raw water or waste water samples quickly.

(iii) Determination of the amount of ionic reagent required. for certain precipitation and neutralization reactions, the end point in such cases is determined by a sudden change in the slope of the conductivity curve..

(iv) An approximate estimation of the eliquots which may be taken for the chemical determination.

(v) Since it is. a measure of dissolved solids it can be used to assess the effect of diverse ions on chemical equilibris, physiological effects on plant or animal, corrosion rates etc.

The standard unit of electrical conductivity is the Siemen per meter (Sm^{-1}) (mhos m^{-1}) in order to avoid the expression of results in small decimal fractions, a smaller unit-the millisiemen per meter (ms m^{-1}) is generally used.

Freshly, prepared distilled water has a conductivity of 0.1 to 0.2 ms m^{-1} (or 1 to 2 umhos cm^{-1}) or even less, but this, increases on storing'due to absorption of atmospheric carbon dioxide to the extent of 0.2 to 0.4 ms m^{-1} or 2 to 4 n mhos cm in about a weeks time. The electrical conductivity of most of the fresh and finished waters is in the range of 5 to 50 ms m^{-1} (or 50 to 500 mhos cm^{-1}). The values for highly mineralized waters go up to l00^rns m^{-4}. and even-higher. Some industrial wastes may have conductivities even more than 10-00 ms m^{-1}.

Electrolytic conductivity increases with temperature at a rate of approximately 2% degree C. Thus, significant errors can result from inaccurate temperature measurements at 25°C.

No chemical preservation of samples is permissible.

(b) Apparatus required – Conductometer:

Many instruments of different ranges are available, most of which are mains operated. Essentially they consist of following the main parts.

(i) *Conductivity cell:* This consists of a pair of rigidly mounted electrodes. Each cell has its own constant depending on its shape and the position and size of the electrodes. This constant is required to calculate the conductivity of the sample. This can be determined by using a standard solution (e.g. 0.01 N KC1 use Table 1 in the conductivity water or by comparison with a .-; cell of accurately known constant).

(ii) *Instrument for measuring electrical conductivity* : Most of the instruments consist of a source of alternating current, a wheatstone bridge and a null indicator. Other instruments measure the relation of alternating current through the cell, a voltage across it and have the advantage of a linear reading of conductance. The instrument chosen should be capable of measuring conductivity with an errors not exceeding 1% of 0.1 Sm^{-1} (or lu mhos cm^{-1}) whichever is greater.

The conductivity cells should be stored in distilled water when not in use. Cells with plantinium electrodes should be repletinized whenever it gives ornatic readings.

To platinize, prepare a solution of 1 g chloropletinic acid (platinium chloride) and 12 mg lead acetate in 100 ml water. Immerce the electrodes in this solution and connect both to the negative terminal of a 1.5 V dry cell battery. Connect the positive side of the battery to a piece of platinium wire and dip the wire into the solution. The amount of current should be such that only a small quantity of gas is evolved. Continue the electrolysis until both cell electrodes are coated with platinium black. The platinizing solution may be saved for subsequest use. The electrodes should be rinsed thoroughly and when not in use keep them, immersed in distilled water.

A different type of electrode with a bright surface is also available. Such cells consists of electrodes made from durable metals like stainless steel are widely used for continuous monitoring and field studies. These cells are calibrated by comparing the conductivity of the water being tested with the results obtained with a laboratory instrument.

Thermometer : Capable of reading to the nearest of 0.1 °C and covering the range 20 to 30°C.

Table 1. Conductance of Potassium chloride solution at 25°C.

Concentration (N)	Conductance (ms m″) Equalant	Specific
0.0001	14,943	14.94
0.001	14.695	147
0.01	14.127	1,413
0.1	12,896	12,980
1.0	11,187	111,900

(c) Reagents required

Conductivity water is used for the preparation of all the standard solutions and this should have a very low conductivity. This must be completely free from carbon dioxide. Reaistilled water must be boiled just before use and allowed to Cool. The conductivity of this water should be less than 0.1 ms m^{-1} .

Standard potassium chloride, 0.01 N : Dissolve 745.6 mg anhydrous KC1 in conductivity water and make upto 1000 ml at 25°C. This is the standard reference solution having specific conductance of 1,41.3 ms^{-1} . It *is* satisfactory for most waters when using a cell with a constant between 1 and 2. For other cell constants, stronger or weaker potassium chloride solution listed in Table 1 will be needed. Store this solution in glass stoppared pyrex bottle.

(d) Procedure adopted

(i) *Determination of cell constant* : Rinse the conductivity cell with at least three portions of 0.01 N KC1 solution. Adjust the temperature of a fourth portion to 25.0 + 0.1°C. Measure the resistance of this portion and note the temperature. Compute the cell corstant C from equation.

(ii) *Conductivity measurements* : Rinse the cell with distilled water followed by portion of sample to be tested. Adjust the temperature of the sample to 25 ± 0.1°C. Measure the resistance of the sample and note the temperature.

(iii) *Calculations:* The electrical conductivity of the sample can be obtained by multiplying the observed conductivity (in ms m^{-1}) by the cell content. If the instrument measure electrical resistance then divide cell constant by observed electrical resistance.

$$E = C = Em \; or \; E = \frac{C}{RM}$$

To account for the temprature differences, divide the result by the factor [1+0.02000 (T - 25)] ic.

$$E = \frac{C \; X \; Em}{1+0.02000(T-25)} \; or \; E = \frac{C}{RM} \frac{1}{1+0.02000(T-25)}$$

Where;

E = Electrical conductivity of sample in ms m^{-1}

C = Cell constant

Em = Electrical conductivity measured ms m^{-1}

T = Sample temperature (°C)

Rm = Electrical resistance measured (K ohm.m)

(e) Expression of results on the basis of experements

Express the result in millisiemen per meter (msm^{-1}) Round off the result as shown below :

Range ms m $^{-1}$	Record to nearest
1.0 to 2.0	0.02
2.0 to 5.0	0.05 .
5.0. to 10.0	0.1
10.0 to 20.0	0.2
20.0 to 50	0.5
0 50 onwards	To nearest whole number

(f) Precision and accuracy maintained

The precision and accuracy of measurement depends on the instrument used and accuracy with which cell constant is determined. The precision and accuracy of about 129? is expected with a satisfactory equipment and a good analyst.

Note : Several conductometers indicate reading as us cm^{-1} or umho cm^{-1} . This reading is to be divided by 10 to convert it int6 msm^{-1}. Following conversions may used :

$$10\ \mu s\ cm^{-1} = 40\ \mu\ mho\ cm^{-1} = msm^{-1}$$

(g) Precautions observed

(i) When electrodes are not being used always keep them in water.

(ii) Always record the temperature and apply temperature correction. Error in this may cause significant error in the results.

(iii) The final conductivity results mainly depend on the cell constant, and hence cell constant determination will cause error in conductivity measurement also.

(iv) Always use conductivity water for preparing standards.

(v) Conductivity measurements should be made on the spot or as soon as possible.

4. Measurment of Hardness

A Hardness in water is mainly due to calcium and magnesium ions. Though some other polyvalent ions such as aluminium, iron, manganese, strontium and *zinc* also contribute to hardness. Li natural waters it is defined as sum of the calcium and magnesium ions expressed as calcium carbonate.

Hardness in water poses problem of scale formation in boilers. It is also objectionable for laundry and domestic purposes since it consumes a larger amount of.soap.

Hardness may range from zero to hundreds of milligrams per liter in terms of calcium carbonate, depending on the source of water.

Estimation of EDTA, Titrimetric Method

(a) Principle adopted

At 2 pH 10+1 EDTA forms a soluble chelated complex with calcium and magnesium. Calcium and complexed magnesium ions also forms a less stable complex with Eriochrome Black T indicator. Thus, when EDTA is added to a flask, metal-indicator complex breaks to form a more stable metal-EDTA complex arid when all the metal is complexed to EDTA indicator shows a sharp change in colour from wine red to blue. Magnesium ion must be present to yield a satisfactory and point in the titration. A small amount of complexometrically neutral magnesium salt of EDTA is therefore added to the buffer. Thus, this step automatically introduces sufficient magnesium and at the same time obviates a black correction.

(b) Interference drawn

Some metal ions interfere in the end point detection which can be overcome by addition of inhibitors. Usually this interference is unlikely in surface waters.

(c) Reagents required

Buffer Solution

(i) Dissolve 16.9 g Ammonium chloride, in 14.3 ml cone. Ammonium hydroxide, add 1.25 g of magnesium salt of EDTA and dilute to 250 ml with distilled water.

(ii) If the magnesium salt of EDTA is not available, dissolve 1.179 g disodium salt of ethylenealimina tetra acetic acid dihydrate (analytical reagent grade) and 780 mg MgSO4. 7H2O or 644 mg MgCl2.6H2O in 50 ml distilled water. Add this solution to 16.9 g NH4C1 and 143 ml cone. NH4OH with mixing and dilute to 250 ml with distilled water.

Store the solution 1 and 2 in a plastic container. Do not store more than a month's supply in a frequently opened container.

(d) Indicator taken

Mix 0.5 g of Eriochrome Black T indicator with 100 g of solid sodium chloride. Grind together in a morter to obtain a homogenious mixture of 40 to 50 mash size. Use about 0.2-0.4 g of the ground mixture for each titration.

Standard EDTA titrant

0.01 N dissolve 3.72 mg analar reagent grade disodium salt EDTA in distilled water, and dilute to 1 litre, standarize against standard calcium. This solution is about 1.00 mg ÇaCOa/ml.

Standard Calcium solution

0.01 M weigh 1.000 g anhydrous calcium carbonate into a 500 ml conical flask. Place a funnel on the neck- of the flask and add slowly 1 N HC1 until all the CaCOs has dissolved. Add 200 ml distilled water and boil for a few minutes to expel out all CO2. Cool, add a few drops of methyl recTindicator and adjust to the intermediate orange colour by adding 3N NaOH or IN HC1, as required. Transfer quantitatively to a one litre volumetric flask and make-up the volume to the mark. This standard solution is equilavant to 1.00 mg CaCOs/ml.

(e) Procedure adopted :

Take 100 ml sample or smaller appropriate portion diluted to 100 ml so that hardness is in range 10 to 20 mg, in a porcelain dish or conical flask. Add 3 ml buffer solution followed by about 0.2-0.4 g of solid indicator. Titrate immediately but slowly with continuous stirring, until the last reddish tinge disappears and blue colour is observed. Reagent blank may be used for comparison. Reagent blank is titrated in a similar way as for the sample.

If the hardness is too low, larger volume may be used and also increase the other reagents and indicator proportionately. Blank may be carried out under identical conditions.

Standarize EDTA titrant with 25 ml calcium standard in a similar way as described for the sample.

(f) Calculations drawn

$$Hardness\ as\ mg/1CaCO_3 = \frac{(V_1 - V_2) X 1000}{ml\ sample} X \frac{25}{V_3}$$

Where,

V1 is the volume of EDTA required for sample.

V2 is the volume of EDTA required for reagent blank.

V3 is the volume of EDTA required for 25 ml standard Calcium.

(g) Expression of Results on the basis of experiments

Hardness is expressed as mg/1 CaCO,3- Round off the results to nearest whole number.

(h) Precision and Accuracy maintained

The precision and accuracy depends on hardness concentration. When the concentration is more than 25 mg/1. precision and accuracy of about ± 2 to 4% can be achieved.

(i) Precautions observed

(i) Carry out the titration within 5 Yninutes to minimize the tendency toward CaCOa precipitation.

(ii) The final volume of sample and blank should be kept constant. Usually 100 ml is better.

(iii) Use deionized or double distilled water for dilutions, blank estimation, etc.

(iv) Use reagent blank for colour comparison.

(v) Add same quantity of indicator for sample and blank.

(vi) Adjust volume of sample such that hardness is in range 5 to 20 mg

5. Estimation of Chloride

Chloride is one of the major inorganic anions in water and waste water. A high concentration occurs in waters from chloridecontaining geological formations. Otherwise, a high chloride content may indicate pollution by sewage or some industrial wastes or an instruction of sea water or other saline water.

The salty taste produced by chloride depends on the chemical composition of the water. A concentration of 250 mg/1 may be detectable in some water containing sodium ions. On the other hand, the typical salty taste may be absent in waters containing 1000 mg/1 of chloride when the predominant cations are calcium and magnesium.

Although chlorides are not harmful, concentration beyond 250 mg/1 impart a peculiar taste to water rendering it unacceptable from aesthetic point of view for drinking purpose. A high chloride concentration harms metellic pipes and structures, as well as agricultural plants.

Two methods, both titrimetric argentrometric and mercurimetric method are described. Wherever colour of the sample interference with the chromate and point in argentrometric method, the other method may be used.

I By Argenometric Method

(a) Principle observed

Chloride is determined in a neutral or slightly alkaline solution by titration with standard silver nitrate, using potassium chromate as indicator. Silver chloride is precipitated quantitatively before red silver chromate is formed.

(b) Interferences drawn

Bromide, iodide and cyanide are measured as equivalent to chloride ion. Sulfide, thiosulphate, cyanide thiocynate and sulfites interfere. They may be oxidised to noninterfering substances as described below :

Take a known volume of sample, add 25 ml of hydrogen peroxide 3% and boil for 15 minutes. Add further 10 ml of hydrogen peroxide and boil for 5 minutes. Test byspotting out for the absence of thiocynate with ferric chloride. Formation of yellowish orange spot indicates presence of thiocyanate. Repeat the heating with hydrogen peroxide till it becomes free from thiocyanate. Orthophosphate in excess of 25 mg/1 interferes by precipitation as silver phosphate. Iron in excess of 10 mg/1 interferes by making the end point.

(c) Reagents required

Silver nitrate standard solution, 0.0282N

Dissolve 4.791 silver nitrate in distilled water and dilute it to 1000 ml. Store in brown glass bottle.

1.0 = 1.0 mg.Cl^-

Standarization titration must be carried out without any dilution.

Potassium chromate indicator solution

Dissolve 50 g K2Cr04 in a little distilled water. Add silver nitrate solution dropwise until definite red precipitation starts. Let it stand for 12 hrs, filter, and dilute to 1 litre with distilled water.

Use chloridefree ie redistilled or deionized distilled water for the preparation of all above reagents. Some water should be used for the sample dilutions also (wherever required). Avoid dilutions as far as possible or apply blank correction.

(d) Procedure adopted

Take 100 ml sample or a suitable portion diluted to 100 ml into a porcelain dish, or conical flask or beaker on a white surface.

Titrate samples in the pH range 7 to 10 directly. Adjust samples not in this range with H2SO4 or NaOH solution. Add 1 ml potassium chromate solution and titrate with silver nitrate solution with constant stirring until only the slightest reddish colouration persists. Be consistent in and point recognition. It is suggested to use reagent and blank as a comparator. If the volume of the titrant exceeds 25 ml then it is better to repeat the method with smaller portion of the sample.

Standarize the silver nitrate titrant with standards sodium chloride. Carry out the blank determination to account for the chloride present in any of the reagents and for the aclubility of silver chromate. A blank of 0.2 to 0.4 is usual for the method.

(e) Calculation drawn

$$Chlorides\, Cl^{-1} = \frac{(V1 - V2)\, X\, 100}{Volume\ of\ sample\, (ml)}\, X\, \frac{25}{3}$$

Where,

V1 = Volume of silver nitrate required for sample

V2 = Volume of silver nitrate required for reagent blank.

V3 = Volume of silver nitrate required for 25 ml of standard sodium chloride.

(f) Expression of results on the basis of experiment

Results are expressed in mgl^{-1} Cl^{-1}

The results should be rounded off as follows :

Concentration results mg I^{-1} less than 20 above 20

Record to nearest mgl^{-1} 0-1 to nearest whole number

(g) Precision and Accuracy maintained

The precision and accuracy of about 2% to 4% is expected from a good analyst, except in the samples were chloride content is less than 50 mg/1.

(h) Precautions observed

(1) Use white opaque background for titration, it is suggested to carry out titrations in white porcelain dish.

(2) The final volume of titrant exceeds 25 ml, it is suggested to take a smaller quantity of the sample.

(3) Also the volume of titrant should not be too small in such cases volume of sample may be increased. It is better to adjust the volume of sample such that titrant value is in range 5-20 ml.

(4) Sample is coloured which may interfere with the chromate end point in such cases mercurimetric method should be used.

(5) Be consistent in endpoint recognition. Reagent blank may be used for comparison.

(6) Use deionized distilled water for the preparation of solution as well as any dilutions.

II Mercuric Nitrate Method

(a)Principle observed

Chloride ions from a highly stable complex with mercuric ions, thus they may be titrated with a standard mercuric nitrate. The end point is detected by diphenylcarbazone indicator in the pH range 2.9 to 3.4 which forms a blue violet complex with an excess of mercuric ions.

(b) Interference drawn

Iodide, bromide, cyanide and thiocyanide are measured as equivalents of chlorides. Sulfide, thiosulphate, sulfite, cyanide and thisulphate interfere seriously with the determination. They may be oxidised to noninterfering substances as described in argentrometric method.

If the sample is highly coloured or turbid to allow the end point to be readily seen this interference may be reduced by the aluminium hydroxide suspension treatment.

Add 3 ml aluminium hydroxide suspension to the measured quantity of sample. Stir thoroughly, lot is settle for few minutes and filter. Wash the precipitate with 10-15 ml distilled water. Also collect washing with the filtrate and continue as described under procedure.

Chromate, ferric, zinc, copper, mangenous and sulfite ions interfere in concentrations above 10 mg/1, aluminium, lead and nickel in concentrations above 100 mg/1. In such cases these titrimetric methods are not suitable.

Phosphate do not interfere.

(c) Reagents required

Mercuric nitrate solution 0.0282N

Dissolve 5.048 g mercuric nitrate, Hg (NO3>2 H2O, in 50 ml water containing 0.5 ml . This solution is approximately 1.0 ml = 1.0 mg Cl^{-1} . Standarize against 25 ml of standard sodium chloride without any dilution. Carry out reagent blank with 25 ml of deionized distilled water.

*Cone. nitric acid dilute to 1 litre and filter if necessary.

Sodium Chloride Standard Solution 0.282N

Dissolve 1.648 g dried sodium chloride in distilled water and dilute to 1 it.

$$1.0 \text{ ml} = 1.0 \text{ mg } Cl^{-1}$$

(d) Indicator required

Dissolve 0.5 g diphenylcarbazone and 0.05 bromophenol blue in 100 ml alcohol. Store in brown bottle.

Nitric acid, approximately 0.05 N

Dilute 3.2 ml cone, nitric acid (density 1.42 g/c.c.) to 1 litre. *Sodium hydroxide, approximately 0.05 N*

2 g sodium hydroxide in distilled water and dilute to 1 litre.

III By Aluminium hydroxide suspension

Dissolve 100 g potassium aluminium sulphate K2SC-4. A12SO4)3- 24 H2O, in about 1 litre of water. Precipitate the aluminium hydroxide by slowly adding ammonium hydroxide until it is in slight excess, with stirring. Allow the precipitate to settle, wash by decantation and check that the final decantate is free from chloride. Dilute the suspension of aluminium hydroxide to about 1 litre.

(1) Procedure adopted

Take 100 ml or smaller portion sothat the chloride content is less than 20 mg, in a porcelain dish. Add 510 drops of indicator. If a blue, violet or red colour develops, add 0.05 N nitric acid drop by drop until it becomes yellow and then add exactly 1 ml in excess. Titrate the solution with standard mercuric nitrate to the point where the first tinge of blueviolet colour appears and which does not disappear on stirring. Prior warning that the end point near is given by the appearance of orange colour.

Determine the reagent blank by titrating 100 ml of distilled water by the same procedure.

(2) *Calculation drawn*

$$Chlorides\ Cl^{-1} = \frac{(V1 - V2)\,X\,100}{Volume\ of\ sample\,(ml)}\,X\,\frac{25}{V_3}\,mg/1$$

Where,

V1 = Volume of mercuric nitrate required for sample

V2 = Volume of mercuric nitrate required for reagent blank.

V3 = Volume of mercuric nitrate required for 25 ml of a standard sodium chloride.

(3) *Expression of results*

Record the results'as mg/1 Cl"

The result should be rounded off as follows :

Concentration range (mg 1")	1-20	20 onwards
Record to nearest (mg 1")	0.1	To nearest whole number.

(4) *Precision and Accuracy*

The precision and accuracy of about 2 to 4% is expected from a good analyst. This also depends on quantity of chloride present, and on the pretreatment necessary.

(5) *Precautions observed*

(i) Titration should be carried out on a opaque white base for which white porcelain dish may be used.

(ii) To final volume should be kept constant say 100 ml.

(iii) The volume of the sample should be selected such that it should not contain more than 20 mg of chloride. On the other hand, volume of sample may be increased if the chloride content is too small.

(iv) Deionized (Chloride free) distilled water should be used for the preparation of solutions as well as for dilution, etc.

6. Estimation of Alkalinity

General : The alkalinity of a water is its quantitative capacity to neutralize a strong acid to a designated pH. Since the alkalinity of many surface water is primarily a function of carbonate, bicarbonate and hydroxide ions, the alkalinity is taken as an indication of the concentration of these constituents. The measured values may also include contributions from borates, phosphates or silikates, if these are present.

The alkalinity of many neutral and treated waters is due only to bicarbonates of calcium and magnesium. The pH of these waters does not exceed 8.3. Their total alkalinity is practically identical with their carbonate hardness and such samples do not have phenolphthelein alkalinity.

Water having a pH above 8.3 contain besides bicarbonates, normal carbonates and possibly hydroxides also. The alkalinity fraction equivalent to the amount of acid required to lower the pH of the sample to 8.3 is called phenolphthalein alkalinity. This fraction is contributed by half of the carbonate and by the hydroxide, if present.

The amount of acid required to lower down the pH to 4.5 determines the total alkalinity or methyl orange alkalinity.

Alkalinity values provide guidance in applying proper losses of chemicals in water and waste water treatment process, particularly in coagulation, softening and operational control of anaerobic digestion. The alkalinity in excess of alkaline earth concentrations is significant in determining the suitability of a water in irrigation.

METHODS ADOPTED

The alkalinity is determined by titration of the sample with a standard solution of a strong mineral acid. Titration is carried out either using indicator or a calibrated pH-meter.

The simple and rapid visual method using an indicator is satisfactory for control and "routine applications. Electrometric method is used for accurate determinations. Electrometric method is used for accurate determinations. It is also free from interference such as colour, turbidity or suspended matters. Low alkalinities (below 20 mg/1 CaCOs) are also best determined by electrometric titration. Thus, for the natural or surface waters were low alkalinity is expected electrometric method is preferred over the visual indicator method.

Wherever possible, the titration should be carried out at the point of sampling. Otherwise, the sampling bottle should be completely filled and the alkalinity is determined within 24 hours.

Interferences drawn

Colour, turbidity and suspended matter may interfere with the visual titration method by masking the colour change of the indicator. Turbidity and suspended matter can be eliminated by simple filtration.

Electrometric method is free from all above interferences.

I By Electrometric Titration Method

(a) *Apparatus required*

Any pH meter utilizing a glass electrode may be used. It must be standardized according to the directions supplied by the manufacturer. It is convenient to use a burrette with a bent jet, sothat the burrete can be used in an upright position.

(b) *Reagents required*

Carbon dioxide free distilled water

It must be used for the preparation of all standard solutions. If the distilled water has a pH lower than 6.0, it should be freshly boiled for 15 minutes and cooled to room temperature.

Sodium carbonate solution 0.05 N

Dilute 3.1 ml cone, sulfuric acid or (10 ml cone, hydrochloric acid) to 1 litre. Standarize against 40.0 ml 0.05N Na2CO3 solution, with about 60 ml water, in a 250 ml beaker by titrating electrometrically to the pH inflection point. This can also be standardized using methyl orange indicator.

Sulfuric acid (or hydrochloric acid) 0.02 N

Dilute 200 ml of 0.1 N standard acid to 1 litre with carbon dioxide free distilled water. Standarize using 10 ml of 0.05 N Na2COa by electrometric or visual titration method. Calculate the normality from equation given below :

$$Normality\ of\ acid, N = \frac{(X)\,X\,(Y)}{53,00\,X\,(Z)}$$

Where,

X is g Na2CO3 dissolved in 1 litre.

Y is ml Na2CO3 taken for titration.

Z is ml standard acid consumed in titration.

When concentration of sulfuric acid is adjusted exactly to 0.1000 and 0.0200 N.

1 ml of 0.1000 N solution = 5.00 mg CaCO3

1 ml of 0.0200 N solution = 1.00 mg CaCO3

(c) Procedure

Take 100 ml sample in a 250 ml beaker, place it on a magnetic stirrer. Dip the electrodes into the solution and adjust the tip of the burette on the surface of the sample. Titrate the standard acid to pH 8.3. Record the volume required, this corresponds to the phenolphthalein alkalinity. Continue the titration to a pH of 4.5. The total titrations value represents the total alkalinity.

II Vfsual Titration Method

(a) Reagents required

Carbon dioxide-free distilled water

Same as described for the electrometric method. *Sodium carbonate 0.05 N*

As described under the previous method. *Phenolphthalein indicator*

Dissolve 0.5 g of phenolphthalein in 50 ml of 95% ethanol and add 50 ml of distilled water. Add to it dilute (e.g. 0.01 N or 0.05 N) carbon dioxide free solution of sodium hydroxide slowly drop by drop till a faint pink colour persists.

Methyl Orange Indicator, Dissolve 0.5 g of methyl orange in 100 ml water.

(b) Procedure adopted

Add 3-4 drops of phenolphthalein indicator to 100 ml of water, in a porcelain dish or a conical flask over a white surface. If no colour is produced, the phenolphthalein alkalinity is zero. If the sample turns pink, titrate with standard acid till pink colour just disappears. In both cases continue the determination using the sample to which phenolphthalein has been added.

Add a few drops of methyl orange indicator, if orange colour is observed without adding any acid, the total alkalinity is zero. If the sample turns yellow, titrate with standard acid until the first appearance of orange colour is noticed.

(c) Calculation observed

$$\begin{bmatrix} Phenolphthalein\ alkalinity \\ (pH\ 8.3)\ mg/1\ as\ CaCO3 \end{bmatrix} = \frac{A\ X\ N\ X\ 50,000}{ml\ of\ sample}$$

$$\begin{bmatrix} Total\ Alkalinity\ (pH\ 4.5) \\ mg/1\ as\ CaCO3 \end{bmatrix} = \frac{B\ X\ N\ X\ 50,000}{ml\ of\ sample}$$

Where,

A is the ml of standard acid required to reach the Phenolphthalein end point or pH 8.3

B is the ml of standard acid required to reach the methyl orange end point or pH 4.5

N is the normality of acid used for the titration

(d) Expression of results on the basis of experiments

Results are expressed as mg/1 of CaCO3. For conversion purposes it may be recalled that 50 mg/1 as CaCO3 are equivalent to 1 m Eq/1.

(e) Precision and Accuracy maintained

No general statement can be made about precision because of the great variation in sample characteristics.

For total alkalinity higher than 50 mg/1, precision and accuracy of electrometric titration are estimated around 1% while for the lower concentrations it may be around 2% to 3%.

The precision and accuracy of visual titration is estimated at 2% for the alkalinity above 50 mg/1 as CaCOs and 2%-W% for alkalinity between 50 mg/1 and 5 mg/1 and 5 mg/1 as CaCOs.

(f) Precautions observed

(i) Do not dilute, concentrate or alter the sample in any way.

(ii) As far as possible also avoid filtration.

(iii) Use only carbon dioxide free distilled water for all purposes such as preparation of reagents, etc.

(iv) If possible titrations should be carried out on the spot. If not the sampling bottle should be filled completely and the alkalinity is determined in shortest possible time, in any case not later than 24 hours.

(v) Use magnetic stirrer for stirring purposes.

(vi) Adjust the volume of sample and concentration of titrant such that the volume of titrant required is appropriate i.e. neither too low nor too high.

(vii) While preparing 0.1 N H2S04, add cone. H2S04 to water (not water to H2SC-4) cool then make up the volume.

7. ESTIMATION OF RESIDUE

A highly mineralized water is less acceptable then a water of little or moderate mineral content for drinking, household and special industrial purposes. A good indication of measuring the quantity of minerals present in water or waste water can be determined by finding out the material left after evaporation.

The term "Residue" applies to the material remaining in the dish after evaporation and drying of a water sample in an oven at a definite temperature. The term "Suspended solids" and "Dissolved Solids" correspond to nonfilterable and filterable residue, respectively, "Fixed Solids" and "Volatile Solids" correspond to the remainder after ignition and loss due to ignition at a definite temperature, respectively.

I Total Residue Calculation

Principle observed : A well-mixed sample is evaporated and dried in a weighted china dish in an oven at 105 + 1°C. The increase in weight represents the total residue.

(1) Apparatus

(a) Evaporating dishes of 150-200 ml capacity made of porcelain, alternatively, beakers can be used.

(b) Drying oven at 105 + 1°C.

(c) Desiccator, provided with a desiccant.

(d) Analytical balance capable of weighing up to 0.1 mg.

(e) Muffle furnace at 550 to 560°C.

(2) Procedure adopted

(i) Heat the clean evaporating dish at 550 ± 5°C for 1 hr in a muffle furnace.

(ii) Cool, desiccate, weight and store the dish in a desicator until it is used. On repetition of this process dish should show a constant weight.

(iii) Transfer a known volume of wellmixed sample to the preweighed dish and evaporate to dryness in the oven at 105 + 1°C. Choose a sample volume in such a way that it will yield a residue between 25 and 250 mg and preferably between 100 and 250 mg (usually, 50 to 100 ml for wastewater and 100 ml to 200 ml for clean waters is suitable).

(iv) After complete evaporation dry it atleast for 1 hr. in the oven.

(v) Cool the dish in a desiccator and weigh it.

(vi) Repeat the cycle of drying, cooling and weighing till the weight of the dish becomes constant.

(3) Calculation done

$$Total\ residue\,(mg/1) = \frac{(X - Y)\,X10^6}{V}$$

Where,

X = Weight of dried sample + dish in g

Y = Weight of empty dish/in g

V = Volume of sample in ml.

(4) Expression of results on the basis of experiments

A report the results as "total residue" or "total solids" in mg/1.

Round off the results to the nearest whole number.

(5) Precision and Accuracy Calculation

The precision is about ± 4 mg or ± 5% for raw waters, the results are considered statistically unreliable due to sampling

II Estimation of Fixed Solids and Volatile Solids

(1) Principle observed

The residue produced on evaporation (at $105 \pm 1^{C}C$ in oven) is ignited in a muffle furnace at 550-560°C. The remainder and the loss on ignition represent fixed and volatile solids, respectively.

(2) Apparatus required

Same as per "Total Residue"

(3) Procedure adopted

(i) After completion of the test for total residue, ignite the dish with the residue in a muffle furnace at 550-560°C for about 30 minutes (30 minutes after muffle furnac? has attained the temperature of 550 560°C).

(ii) Allow the dish to cool partially in air most of the heat has dissipated or switch off the muffle furnace and let the temperature fall.

(iii) Transfer the dish to a desiccator, cool and weigh it as soon it has cooled, completely.

(iv) Repeat the operation of ignition, cooling and weighing until a constant w.eight is obtained (i.e. till the difference between two successive series of operation is not more than 0.5 mg or 1% of the remainder solids).

(4) Calculation derived

$$Voilatile\ Residue, mg/1 = \frac{(X - Y)X10^6}{V}$$

$$Fixed\ Residue, mg/1 = \frac{(Y - Z)X10^6}{V}$$

Where,

X = Weight of sample + dish, before ignition but after evaporation in g.

Y = Weight of sample + dish after ignition in g.

Z = Weight of dish in g.

V = Volume of sample in ml.

(5) Expression of Results on the basis of experiments

Report the results as "fixed residue" or "fixed solids" and "Volatile residue" or "volatile solids"in mg/1. Round off the results to the nearest whole number.

(6) Precision and Accuracy Calculation

Same as per "Total residue".

III Dissolved Solids material

(1) Principle

Dissolved solids or filterable residue, are those solids capable of passing though a standard glass fibre filter paper. (In waters containing no suspended solids, the dissolved solids are equal to the total solids). The filtrate from the nonfilterable residue or suspended solids can be used to determine the filterable residue.

Fixed dissolved solids and "Volatile dissolved solids" correspond to the filterable solids remainder after ignition and loss by ignition, respectively.

(2) Apparatus required

(i) Glass-fibre filter paper, Whatman GF/C grade or equivalent.

(ii) Millipore filtration assembly with suction flask of 500 ml or 1000 ml capacity.

(iii) Porcalein dish or beaker, 150-200 ml capacity

(iv) Drying oven, 180°C ± 2°C or 105°C ± 1°

(v) Desiccator

(vi) Analytical balance capable of weighing upto 0.1 mg.

(3) Procedure adopted

(i) *Preparation of evaporating dish :* Heat the clean dish to 550-560°C for about one hour in a muffle furnace. Cool, partially in air then in desiccator. Weigh immediately before use.

(ii) *Preparation of glassfibre disc :* Place the disc on the membrane filter holder. Wash it with three successive 20' ml volume of distilled water. Discard all these washings.

(iii) Under vacuum filter 100 ml or more of well mixed same through glass fibre filter.

(iv) Transfer 100 ml (or a larger volume if total filterable residue is low) filtrate to a preweighed evaporating dish.

(v) (a) *Dissolved Solids at 105°C :* Evaporate and dry the sample in a dry oven at 105 + 1°C (Dry atleast for 1 hr after complete evaporation). Cool in a desiccator and weigh.

Repeat the drying cycle until a constant weight is obtained. (i.e. ± 0.5 mg or ± 1% of the evaporated sample).

(b) *Dissolved Solids at 180°C :* The above procedure to be carried out at 180 ± 2°C instead of 105 ± 1°C.

(4) Calculation direved

$$Total\ disslove\ solids\ (mg/1) = \frac{(X - Y) X 10^6}{V}$$

Where,

X = Weight of dried filtered sample + dish in g

Y = Weight of empty dish in g.

V = Volume of sample in ml.

(5) Expression of results on the basis of esperiments

Express the results as "Total dissolved solids" dried at -°C, mg/1.

IV Estimation of Volatile and Fixed Dissolved Solids

(1) Procedure

1. Determine the proportion of matter volatile on ignition by igniting the dried residue (after during dissolved solids) in a muffle furnace at 550°C for about an hour.

2. Cool in a desiccator and weigh it.

(2) Calculations derived

$$Volatile\ disslove\ solids\ (mg/1) = \frac{(X - Y)\,X\,10^6}{V}$$

Where,

X = Weight of dried residue + dish, g (before ignition)

Y = Weight of ignited residue + dishg (after ignition)

V = Volume of filtrate taken ml.

$$Fixed\ disslove\ solids, mg/1 = \frac{(Y - Z)\,X\,10^6}{V}$$

Where,

Y = Weight of ignited residue + dish, g

Z = Weight of dish, g

V = Volume of filtrate taken ml.

(3) Expression of results on the basis of experiments

Express results as "Volatile dissolved solids" and "Fixed dissolved solids" as mg/1. Round off the results to the nearest whole number.

V Estimation of Suspended Solids

(1) Total suspended solids

Suspended solids are those solids which are retained on a standard glass fibre filter disc.

(2) Samples handling and preservation

Non-homogenous particulate matter such as leaves, sticks, fish and lumps of focal matter should be excluded from the sample.

Preservation of the sample is not practical, analysis about be carried as soon as possible.

(3) Apparatus required

(i) Glassfibre filter disc, Whatman GF/C grade or equivalent, 70 mm

(ii) Millipore filtration assembly with suction flask

(iii) Drying oven at 105 ± 1'C

(iv) Desiccator

(v) Analytical balance capable of weighing to 0.1 mg

(4) Procedure adopted

Preparation of glass fibre filter disc :

(a) After numbering the filter disc dry it in an oven at 105 ± 1°C for about an hour. Cool in a desiccator and weigh it.

(b) After weighing place it on the filter holder. Apply vacuum and wash it with about 20 ml of distilled water.

While the pump is on transfer 100 ml or an appropriate of volume of the wellmixed sample so that it contains about 25 to 200 mg ms of suspended solids, to the filtration of the milipore assembly. If the solids are very high add small volume in number of installments.

After complete filtration of the sample add about 20 ml distilled water.

Transfer the filter paper carefully to the petridish when it is complete dry. Dry at least one hour at 105 ± 1°C in the oven. Cool in a desiccator and weigh.

Repeat the drying cycle until a constant weight is obtained.

(5) Calculation derived

$$Total\ suspended\ solids\,(mg/1) = \frac{(X-Y)\,X10^6}{V}$$

Where,

X = weight of filter paper + residue, g

Y = weight of filter paper, g

V = volume of sample, ml

(6) Expression of results on the basis of experiments

Express the results as total suspended solids in mg/1. Round off to the nearest whole number.

VI Volatile and Fixed suspended solids

(i) Procedure adopted

(1) After determining the total suspended solids the filter disc containing the suspended matter is ignited in the muffle furnace at 550°C for about an hour.

(2) Cool in a desiccator and weigh.

(2) Calculation derived

$$Volatile\ suspended\ solids\,(mg/1) = \frac{(X-Y)\,X10^6}{V}$$

Where,

X = weight of dried residue + filter disc, g (before ignition)

Y = weight of ignited residue + filter disc, g (after ignition)

V = volume of sample filtered, ml.

$$Fixed\ suspended\ solids, mg/1 = \frac{(Y-Z)\,X10^6}{V}$$

Where,

Y = weight of ignited residue + filter disc, g

Z = weight of filter disc, g
V = volume of sample filtered, ml

(3) Expression of results on the basis of experiments

Express the results as volatile suspended solids and fixed suspended solids as mg/1. Round off to value the nearest whole number.

(4)General Precautions followed

(a) Always set the balance to zero befcnftuse.

(b) Before weighing ensure that dish of filter disc is cooled to the room temperature.

(c) At all stages prevent dish or filter disc from dust, etc.

(d) Greater the number of replicates more is the accuracy. Keep at least 3 replicates.

(e) Mark the number on the dish or filter disc before altering the test (before drying).

8. Estimation of Dissolved Oxygen (Do)

(1) General Principle

TiDissolved oxygen (DO) in natural and waste water depends on the physical, chemical and biological activity in *a* water body. The analysis for DO is important in water pollution control and waste treatment process control. DO as a parameter used :

(i) To asses quality of surfacewater and to keep a check on stream pollution.

(ii) It is the basis for BOD which is an important parameter to evaluate pollution potential of wastes.

(iii) It is responsible for corrosivity, photosynthetic activity and septicity.

(iv) DO is necessary for all aerobic biological wastewater treatment processes.

The iodometric test is precise and reliable titrimetric procedure for DO analysis. The test is based on the addition of divalent manganese solution, followed by strong alkali, to the sample in a BOD glass stoppared bottle. DO present, rapidly oxidizes an equivalent amount of the dispersed divalent manganous hydroxide of higher oxidation state (brownish-orange in colour). This brownish-orange colour indicates presence of oxygen. In the presence of iodide and on subsequent acidification, higher manganese hydroxide revert to the divalent state, and liberate iodine equivalent to the original dissolved oxygen content of the sample. The iodine liberated is titrated with a standard solution of sodium thisulphate using starch as an indicator.

(2) *Interference drawn*

It is caused by certain oxidizing agents which liberate iodine from iodides. (Such as ozone, chlorine, ferric compounds, manganese in tri or higher oxidation states chromate, nitrate, persulphate, peroxides, etc. by certain reducing agents which may reduce iodine to iodide (Such as Ferrous compounds, thiosiilphate, readily oxidizable organic matter, nitrite etc.). Some organic compounds hinder the setting of the precipitate and colour of some may interfere in the starch end point. Modifications are included for minimize the effect of interfering materials.

(3) *Apparatus required*

The bottle used should be of good quality having 300 ml capacity with narrow neck and well fitting ground glass stopper. It is convenient to have number on bottles and their stoppers.

(4) *Reagents required*

(i) *Mangenous sulphate solution* : Dissolve 500 g MnS04 in distilled water. Filter and dilute to 1 litre. This mangenous sulphate solution should not give a colour with starch when added to an acidified solution of potassium iodide.

(ii) *Alkali-iodideazide reagent:* Dissolve 500 g NaOH (or 700 g KOH), and 135 g Nal, (or 150 g KI) in distilled water and dilute to 1 litre. Add 10 g sodium azide, NaNa dissolved in 40 ml distilled water. Potassium and sodium salts may be used interchangeably. This reagent should not give colour with starch solution when diluted and acidified.

(iii) *Sulfuric acid* : H2SO4 cone, one milliliter of this is equivalent to about 3 ml of alkaliiodideazide reagent.

(iv) *Starch indicator:* Make a smooth paste of 5 g soluble starch in cold water and pour this into 1 litre boiling water with constant stirring. Boil for one minute and allow to cool before use. Use clear supernatant. Preserve with 1.25 g salicylic acid /I or by adding a few drops of toluene.

(v) *Sodium thiosulphate stock solution 0.125 N* : Dissolve 31.5 g sodium thiosulphate pentahycfrate, Na2S2O3, 5H-2O in 1 Htre copper free boiled and cooled distilled water. Add 5ml chloroform or Ig NaOH as preservative.

(vi) *Standard sodium thiosulphate titrant 0.125 N* : Dilute the stock sodium thiosulphate solution to obtain 0.0125 (100 ml of stock 0.125 M Na2S2O3 diluted to 1000 ml with distilled water gives 0.0125 N Na2S2O3) standarize this diluted solution daily.

Standarization

Pipette 10 ml 0.012 5N Potassium iodate into a conical flask containing about 100 ml distilled water. Add 2 ml of cone. H2SO4 followed by about 2 g of potassium iodide solid. Titrate immediately against sodium thiosulphate 10/25N using starch as indicator. Apply concentration correction in the calculation.

Potassium iodate, 0.0125 N

Dissolve 0.146 potassium iodate previously dried at about 120°C, in distilled water and dilute to 1 litre. The solution is stable for a long period if stored in glassstoppared bottles.

(5) Procedure observed

(i) Add 2 ml mangenous sulphate followed by 2 ml alkali-iodide-azide reagent, dipping thc pipctte little below the surface.

(ii) Stopper carefully to exclude air bubbles and thoroughly mix the contents. To attain this invert the bottle at least 15 times.

(iii) Allow the precipitate to settle to the lower one third of the bottle, repeat the mixing and allow the precipitate to settle completely leaving a clear supernatant liquid.

(iv) Add '2 ml cone. H2S04 immediately after removing the stopper. Restopper, and mix by gentle inversion until all the precipitate dissolves, if it does not, allow to stand for a few minutes and repeat the mixing.

(v) Distribute the iodine uniformly before taking a part of it titration.

(vi) Take 100 ml of sample for titration (1.3 ml extra to account for 4 ml or the reagents, mangenous sulphate and alkali-iodide-azide, added to 300 ml of sample) in a conical flask and immediately titrate the liberated iodine with the standard thiosulphate solution to a pale yellow straw colour. Add 2 ml starch solution and continue the titration till the first disappearance of the blue colour. Disregard subsequent recolouration due to the catalytic effect of nitrite of ferric salts.

It must be remembered that iodine is volatile and therefore the titration must be carried out as expediously as possible and with the minimum of exposure to the air.

MODIFICATION OF RIDEAL-STEWART FOR ELIMINATION OF INTERFERENCE BY FERROUS IRON

This modification is used when the sample contains ferrous iron. Add 0.7 ml cone. H2S04 followed by 1 ml potassium permangnate (0.63%) and 1 ml potassium floride (40%) to the sample bottle, stopper and mix by inversion. The amount of permanganate added should be just sufficient to obtain a violet tinge that persists for 5 minutes. Remove the excess permanganate by adding 0.5 to 1.0 ml potassium oxalate (2%). Mix well and let stand in the dark to fecilitate the reaction.

Excess oxalate causes low results and hence add only an .amount of oxalate that completely decolourizes the potassium permanganate.

After applying this modification usual method should be followed.

(6) Calculation derived

$$\begin{bmatrix} \textit{Dissolve oxygen} \\ \textit{Content mg/1} \end{bmatrix} = \frac{\begin{matrix} \textit{Volume of thiosulphate} \\ \textit{required for } 101.3\, ml \textit{ sample } X\, 10 \end{matrix}}{\begin{matrix} \textit{Volume of thisoulphate required for} \\ 10\, ml \textit{ of } 0.0125\, N \textit{ Potassium iodate} \end{matrix}}$$

(7) Expression of results on the basis of experimets

Express the dissolved oxygen results in mg/1. Report the values to the nearest of first decimal i.e. 0.1 mg/1.

(8) Precision and Accuracy observed

The DO can be measured with a precision of about 50 ug/1 or 0.05 mg/1 in surface and slightly polluted waters, expressed as standard deviation. In the presence of appreciable interference this may be as high as 100 ug/1 or 0.1 mg/1.

(9) Precautions followed

(i) Do not use a cap or synthetic detergents for washing the DO bottles. Normally bottles are kept clean by the acidic iodine solution of the winkler procedure and require no further treatment except thorough rinsing with the tap water.

New bottles should be cleaner with 5N sulfuric or hydrochloric acid and then rinsed thoroughly with the tap water.

(ii) Just before taking the sample rinse sample bottle with the sample.

(iii) Bottle should be completely filled with the sample and remove bubbles entrapped into the bottle while taking sample by tapping the neck of the bottle with the stopper.

(iv) Fix the DO on the site by adding manganous sulphate and alkaliiodideazide solution.

(v) The titrations should be carried out as expediously as possible and with minimum exposure to the air, so as to avoid any losses of iodine during titration.

(vi) Use BOD samples discharger to avoid inconvenience caused in measuring 101.3 ml. This will save time as well as maintain uniformity'in the comparative analysis.

(vii) Avoid errors due to carelessness in collecting the samples, prolonging the completion of the test or due to any interferences.

9. Estimationof Biochemical Oxygen Demand (Bod)

(1) General Principle

Biochemical oxygen demand (BOD) is defined as the amount of oxygen required by micro organisms to stablize biological decomposable organic matter in a waste under aerobic conditions. The biochemical oxygen demand is an approximately measure of the amount of biochemically degradable organic matter present in the sample.

The BOD test is widely used to determine (1) the degree of pollution in lakes and streams at any time and their self purification capacity, (2) the pollutional load of waste waters and (3) efficiency of waste treatment plants.

Since the test is based on mainly bioassay procedure measuring amount of oxygen consumed by bacteria while stabilizing organic matter under aerobic conditions, it is necessary to provide standard conditions of nutrient supply, pH, absence of substances inhibiting microbial growth and temperature. Because of the low solubility of oxygen in water, the sample is diluted proportionately depending upon the expected BOD sothat the demand does not exceed the available oxygen. A mixed group of organisms should be present in the sample, if not, the sample has to be seeded artificially. Temperature is controlled at 20°C. The test is carried out for 5 days, as 70 to 80'/r of the BOD is satisfied during this period.

(2) Interferences and Inadequacies drawn.

The pH of the sample should be adjusted by adding alkali or acid, in the range 6.5 to 8.5.

Some samples may be sterile, and will need seeding. The purpose of seeding is to introduce into the sample a biological population capable of oxidizing the organic 'matter. Where such microorganisms are already present as in surface waters, domestic sewage or unchlorinated effluents etc., seeding is not required.

For seeding, wherever necessary use settled as to each litre of dilution water.

Other interference in BOD test are the same as in the DO test. In addition, lack of nutrients in dilution water, lack of acclimated seed organisms and presence of heavy metals or other toxic materials such as residual chlorine are other sources of interference in this test.

(3) Apparatus required

Incubation bottles : 300 ml capacity with groundglass stoppers. New bottles should be cleaned with either 5 N hydrochloric or sulfuric acid followed by rinsing with distilled water. In normal use bottles once used for winkler procedures does not required any treatment apart from thorough rinsing with tap water and distilled water. As a precaution against air into the dilution bottle during incubation, use special 800 bottles and add water to the flared mouth of the bottle.

(4) Reagents required

Dilution water : Distilled water alone is unsatisfactory as a diluent and it is recommended that a synthetic dilution water be employed. This can be prepared by adding reagents to good quality of distilled water. Water from copper stills should not'be used since copper inhibits biochemical oxidation. Glass distilled or deionized water produced in some commercial units may be used.

Stocks solutions of the follow ing pure chemicals are required. Any solutions showing signs of precipitates or growth should be discarded.

Phosphate buffer stock solution : Dissolve 42.5 g potassium dihydrogen phosphate (KH2PO4) in 700 ml water and add 8.6 g sodium hydroxide. This should give a solution of pH 7.2 which should be checked. Add 2 g ammonium sulphate and dilute to 1 litre.

Magnesium sulfate solution : Dissolve 25 g magnesium sulfate MgSO4.7H20 in 1 litre water.

Calcium chloride solution : Dissolve 27.5 g anhydrous calcium chloride CaCl2 in 1 litre distilled water.

Ferric- chloride solution : Dissolve 0.125 g ferric chloride FeCl3.6H2O in 1 litre water:

(5) The Seeding

The purpose of seeding is to introduce into the sample a biological population capable of oxidizing the organic matter in the waste water. Samples in which such microorganisms are already present, as in domestic sewage and surface waters seeding is not required.

When the sample contains very few microorganisms the standard seed material used is sewage either that has been stored at 20°C for 24 to 36 hrs. The seed added should produce a seed correction of atleast 0.6 mg/1. Usually 2 ml to 5 ml of sewage is used per litre of dilution water.

(6) Procedure observed

Method Preparation of dilution water

Aerate the required volume of distilled water in a container by bubbling compressed air for 12 days to attain DO saturation. After aeration keep it at 20°C for atleast on day.

At the time of use, add 1 ml each of phosphate buffer, magnesium sulphate, calcium chloride and ferric chloride for each litre of dilution water. The trend of adding these nutrients should be followed as given above. If required add requisite uantity of seed also. Mix well.

Method Dilution of sample

Neutrialize the sample to pH around 7.0 using alkali or acid (NaOH or $H2S0_4$).

The sample should be free from residual chlorine. If it contains residual chlorine, remove it by using Na2SOs solution as given below :

Take 50 ml of the sample and acidify with 10 ml of 1 + 1 acetic acid. Add about 1 g of KI. Titrate with Sodium sulfite using starch as indicator. Calculate the volume of Na2SOa required per ml of the sample and add accordingly to thc sample to be tested for BOD.

Atleast two dilutions of the sample are made so that the depletions are in the range 40 to *70%*. It is recommended to have lower dilution t.o give *40%* and higher one to give *70%* oxygen depletion. The dilutions may be made as follows :

Requisite quantity of the sample is taken in a one litre capacity volumetric flask. Dilute this by siphoning out the dilution water to the mark. Mix well. Rinse BOD bottles with the sample and then fill the bottles with the diluted sample. As for as possible avoid entrapping of air bubbles in the bottle. Tap the neck of the bottles to ensure the removal of the air bubbles from the bottle. Stopper the bottle immediately.

Keep one bottle for determination of initial (zero day) dissolved oxygen and incubate two bottles at 20°C for 5 days. The bottles incubated should be water sealed. The'water to seal the bottle should be added daily, as some water is lost due to evaporation.

Prepare four blanks by siphoning out only dilution water into the bottle. These bottles should be first rinsed and then filled with the dilution water. Two of these blanks should be used to determine initial dissolved oxygen and rest of the two are incubated for 5 days at 20°C. Usually depletion due to blank in 5 days should not be more than 0.2 mg/1.

Dissolved oxygen, initial and after five days is determined as described in the chapter for. the determination of the dissolved oxygen.

Seed correction : If the dilution water is seeded, do not use the seeded blank for seed correction because the 5day seeded dilution water blank is subject to aeratic oxidation due to the very high dilution of seed, which is not the oxygen depletion of the seed by setting up a separate series of dilutions and selecting those resulting in 40 to 70% oxygen depletions in 5 days. Use one of these or average of these or average of these depletions to calculate the correction due to the small amount of seed in the dilution water.

(7) Calculations derived

(i) When BOD has been determined in an undiluted sample:

BOD mg/1 = DO before incubation mg/1 – DO after incubation mg/1.

(ii) When BOD has been determined in a diluted sample :

Definitions –

Dl = DO of diluted sample on Oth day
D2 = DO of diluted sample on 5th day
B1 = DO of dilution blank on Oth day
B2 = DO of dilution blank on 5th day
Cl = DO of seed control on Oth day
C2 = DO of seed control on 5th day

(iii) F = Ratio of seed in sample to seed in control

$$F = \frac{\%\, seed\ in\ D1}{\%\, seed\ in\ C1}$$

Seed correction = (C1 - C2) F

(iv) BOD without applying seed correction

$$BOD\, mg/1 = \frac{(D1 - D2) - (B1 - B2)\, X\, 100}{\%\, sample}$$

(v) BOD applying seed corection

$$BOD\,mg/1 = \frac{(D1-D2)-(C1-C2)\,X\,100}{\%\,sample}$$

Volume correction

This correction is applied in case of low BOD samples were longer volume of sample is incubated. In such cases the fraction of dilution water used to dilute the sample is much less than 100% and hence depletion due. to blank, for the added fraction only should be" taken into account and not the 100% in normal cases.

The following modified equation is to be used for applying this correction :

(A) BOD without applying seed correction

$$BOD\,mg/1 = \frac{(D1-D2)-(B1-B2)\,X\,100}{\%\,sample}$$

(B) BOD without applying seed correction

$$BOD\,mg/1 = \frac{(D1-D2)-(C1-C2)\,X\,100}{\%\,sample}$$

Where A is the fraction of dilution water added and is given as

$$A = \frac{100-\%\,Sample\,added}{100}$$

(8) Expression of results

BOD is expressed as mg/1, 5 days 20°C .

Precision and Accuracy : There is no acceptable procedure for determining the accuracy of the BOD test.

(9) Precautions followed

(i) The trend of adding nutrients should be phosphate buffer, magnesium sulphate, calcium chloride and ferric chloride.

(ii) After aerating dilution water stablize it at 20°C for atleast one day.

(iii) The pH of the sample should be in range 6.5 to 8.5.

(iv) While siphoning the dilution water and filling BOD bottles avoid entrapping of bubbles.

(v) Wherever necessary apply seed and volume corrections.

(vi) Add water on the neck of the bottles during incubation period to ensure continuous proper water sealing.

(vii) Requisite treatment should be given to the samples containing interfering radicals like iron, residual chlorine, etc.

10. Estimation of Chemical Oxygen Demand (COD)

(1) General Principle

The chemical oxygen demand determines the amount of oxygen required for chemical oxidation of organic matter using a strong chemical oxidant, such as potassium dichromate under reflux conditions.

The test is widely used to determine :

(i) the degree of pollution in water bodies and their self purification capacity.

(ii) efficiency of treatment plants

(iii) pollutional loads

(iv) provides rough idea of BOD, which can be used for BOD estimation.

The limitation of the test lies in its inability to differentiate between the biologically oxidizable and biologically inert material.

COD determination has an advantage over BOD test in that the results can be obtained in less than 5 hours where as BOD determination requires 5 day. Further the test is relatively easy and precised. Also there are not many interferences as in the case of BOD.

(2) Principle observed

Most of the organic matter are destroyed when boiled with a mixture of potassium dichromate and sulfuric acid producing carbon dioxide and water. A sample is refluxed with a known amount of potassium dichromate in sulfuric acid medium and the excess of dichromate is titrated against ferrous ammonium sulfate. The amount of dichromate consumed is proportional to the oxygen required to oxidize the organic matter.

(3) Interference drown

Straightchain aliphatic compounds, aromatic hydrocarbons, fatty acids, chlorides, nitrites and iron are the main interfering radicals.

The interference caused by chlorides can be eliminated by the addition of mercuric sulfate to the sample prior to addition of other reagents. About 0.48 of HgS04 is adequate to complex 40 mg Cl ions in the form of poorly ionized HgCh.

Nitrate exerts a COD of 1.1 mg/mg N. 120 mg of sulphuric acid is added to K2Cr2O7 solution to avoid interference caused by NOa. When 20 ml sample and 10 ml dichromate is taken, this can take care of concentrations up to 6 mg/1 NO~2N.

Slider sulfate is added to cone. H2SO4 (22g/4kg acid) as a catalyst. This accelerates the oxidation of straight chain aliphatic and aromatic compounds.

For complete and better oxidation of organic matter it is necessary to see that the final concentration of H2SO4 cocn. is 50%.

(4) Apparatus required

Reflux Apparatus consisting of a flat bottom, 250 to 500 ml capacity flask with ground glass joint and a condenser with 24/40 joint.

Hot plate : It is convenient to have a bigger hot plate on which a series of refluxing sets is connected to have a single water circulation to all. This also eliminates the botheration of having separate blank for each sample.

(5) Reagents required

Standard Potassium dichromate, 0.25 N : Dissolve 12.259 g potassium dichromate previously dried at 103°C for 24 hrs in distilled water. Add about 120 mg sulpemic acid to this. Dilute to 1 litre.

Sulpemic acid : Add 22 g silver sulfate to 9 Ibs (4kg) cone. H2SO4 bottle. Keep overnight for dissolution. Shake well after dissolution.

Standard Ferrous Ammonium Sulfate, 0.1 N

Dissolve 39 g Fe(NH4)2(SO4)2.6H2O is distilled water. Add 20 ml cone. H2S04, cool, and dilute to 1 litre. Stahdarize this solution daiW against the standard

Technique for Standardization

Dilute 10.00 ml standard K2Cr2O7 solution to about 100 ml. Add 30 ml cone. H2SO4 and cool. Add 3-4 drops of ferroin indicator, and titrate with ferrous ammonium sulfate till the colour changes to wine red.

$$\begin{bmatrix} Normality\ of\ Ferrous \\ Ammonium\ Sulfate \end{bmatrix} = \frac{10\,X\,0.25}{ml\ Fe(NH_4)_2\,(SO_4)_2}$$

Ferroinindicator

Dissolve 1.485 g 1,1-Dpheninthroline monohydrate, and 695 mg FeSO4.7H2O in water and dilute to 100 ml.

This indicator solution may be purchased already prepared. Mercuric Sulfate, HgSO4 Crystals

(6) Procedure adopted

Place 0.4 g HgSO4 in a reflux flask. Add 20 ml or an aliquot of sample diluted to 20 ml with distilled water. Mix well, sothat chlorides are converted into poorly ionized mercuric chloride. Add 10 ml standard K2Cr2O7 followed by slowly 30 ml sulfuric acid which already contains silver sulfate. This follow addition alongwith swirling prevents loss of volatile materials such as fatty acids, in the sample. Mix well if the colour turns green either take fresh sample with smaller aliquot or add more dichromate and acid. Final concentration of cone. H2SO4 should be always 50%.

Connect the flask to condensers and reflux for 2 hours. Cool and wash down the condensers with small quantity of distilled water. Remove the flask and add about 80 ml distilled water. Cool and titrate against standard ferrous ammonium sulfate using ferroin as indicator. Colour changes sharply from green-blue to wine red.

Reflux a reagent blank under identical conditions preferably simultaneously with the sample.

Alcornate procedure for low-COD samples (below 50 mg/1)

The method is same as for the high-CO.D samples except in this case the concentrations of K2Cr2O7 and ferrous ammonium sulfate is reduced i.e. K2Cr2O7, 0.05N and ferrous ammonium sulfate, 0.025N. Exercise extreme care with this procedure because even a trace of organic matter in the glassware or the atmosphere may cause a gross error. COD values upto 160 mg/1 can be estimated by this procedure.

(7) Calculation direved

$$mg/1\,COD = \frac{(a-b)\,N\,X\,8000}{ml\ sample}$$

Where,

a is the ml Fe(NH4)2 (804)2 required for blank
b is the ml Fe(NH4)2 (804)2 required for sample
N is the normality of Fe(NHU)2 (804)2

(8) Expression of Results on the basis of experiments

Results are expressed as mg/1 of O2

Report to the nearest whole number.

(9) Precision and Accuracy

Precision and accuracy both depends upon the COD value.

For the sample having COD higher than 400 mg/1 precision upto 2% is expected from .a good analyst. As the COD value goes on decreasing precision also become proper and poorer i.e. % goes on increasing.

Precision for the low COD samples may be improved by using alternate method where diluted reagents are used.

The accuracy of this method is till not very clear. For most organic compounds the oxidation is 95 to 100% of the theoretical value.

(10) Precautions followed

(i) A strength of the sulfuric acid in final solution should be 18N.

(ii) Trend of making mixture should be $HgS0_4$, sample (swirl), $K_2Cr_2O_7$ and cone. H_2SO_4 (slowly with swirling).

(iii) As for as possible reflux blank and samples simultaneously.

(iv) After refluxing, cool, use required amount of distilled water for washing the condenser. Cool and then titrate.

11. Estimation of Nitrogen

A Nitrogen content in sewage water can be determined by nesselarization method as suggested by Me. Donald (1978).

Procedure adopted

One ml of sewage water is taken for digestion in corning test tube which is digested with few drops of cone. H2S04 and a pinch of digestion mixture (The digestion mixture consist of potassium sulfate, sodium triosulfate, salicylic acid and selenium), after cooling few drops of 30% hydrogen peroxide is added and heating is continued for some time. Finally the colourless fluid is made up with distilled water to volume 5 ml, and to this 0.2 ml of sodium silicate (10%) and sodium hydroxide (10%) are added and shake thoroughly, the final volume is made to 10 ml, with distilled. Then, 2 ml of diluted (2:1) nesseler's reagent is mixed and shaked. The absorbance of yellow colour after 10 minutes is read at 425 nm. by spectrophotometer. The amount of nitrogen in samples is calculated with the help of standard curve, prepared from ammonium sulfate and results are expressed as ug nitrogen/ ml of sewage water.

12. Estimation of Sulfate

(1) Generay Indtroduction

Sulfate is widely distributed in nature and may be present in natural waters in concentrations ranging from a few to several thousand milligrams per litre. Mine drainage wastes may contribute high sulfate by virtue of pyrite oxidation. Sodium and magnesium sulfate exert cathatic action and hence its concentration above 250 mg/1 in potable water is objectionable. Sulfates cause a problem of scaling in industrial water supplies, and problem of odour and corrosion in waste water treatment due to its reduction to H2S.

Though the gravimetric method is most accurate for sulfate concentration above 10 mg/1, natural waters are likely to have low concentrations and hence turbidimetric method which is more rapid and also is more or less equally accurate as the former method, is suggested. Turbidimetric method may be applied to the concentrations up to 60 mg/1. When the concentration exceeds 60 mg/1 it may be diluted accordingly.

(2) Principle direvid

Sulfate ions are precipitated in a hydrochloric acid medium with barium chloride so as to form a uniform suspension of BeS04 crystals. The absorbance of the suspension is measured by a photometer and the sulfate concentration is determined by comparison of the reading with a standard curve.

(3) Interference drown

Colour and suspended matter interfere in determination. Some suspended matter may be removed by filtration. If both are small in comparison to sulfate on concentration, this interference is corrected by running blanks in which BaCl2 is eliminated. Silica in excess of 500 mg/1 will interfere.

There are no ions other than sulfate in normal waters that will form insoluble compounds with barium under strongly acid conditions.

Minimum detectable concentration : Approximately 1 mg/1 sulfate.

(4) Apparatus required

(i) *Magnetic Stirrer* : It is better to use a magnetic stirror to obtain constant stirring rate and time (usually one minute is sufficient).

(ii) *Photometer* : One of the following is required.

(a) Nelholometer
(b) Spectrophotometer (at 420 nm)
(c) Filter photometer (at 420 nm, violet blue filter)

(iii) Stopwatch.

(5) Reagents required

Conditioning reagent : Mix 50 ml glycerol with a solution containing 30 ml Cone. HC1, 300 ml distilled water, 100 ml 95% ethyl alcohol and 75 g NaCl.

Barium Chloride BaCh : Crystals, 20-30 mash.

Standard Sulfate Solution : (a) Dissolve 1.4798 anhydrous Na2SO4 in distilled water and dilute to 1000 ml.

Dilute 100 ml of solution "a" to 1 litre. This solution is 1 ml = 100/ue SO_4.

(6) Procedure adopted

Take 100 ml conical flask. Add exactly 5 ml conditioning reagent and mix it on the magnetic stirror. Add a spoonful of BaCl2 crystals while stirring. Continue stirring for exactly 1 min. at a constant speed.

Immediately after the stirring is over, four some samples into the absorption cell of the photometer and measure the turbidity after every 30 sec. for 4 minutes at 420 nm (violet blue filter). Maximum turbidity usually occurs within 2 min. and the reading remains constant thereafter for 3 to 10 min.

Prepare standard curve by carrying standard sulfate solution through entire procedure. Space standards at 6 mgA increment in the range 0 to 40 mgA. Run reagent blank under identical conditions, it is convenient to adjust the photometer (either 100% transmission or zero absorbance) with this blank before taking the sample readings.

To account for the interference due to the colour and turbidity of the sample, run sample blank in which only BaCl2 is withstand. For a particular sample either this may be used to adjust the instrument or the absorbance due to this is deducted from the sample absorbance. Use same amount of sample for the sample blank and sample determination.

(7) Calculation derived

$$mg/1 SO_4 = \frac{A X 1000}{ml\ sample}$$

Where A is the concentration of sulfate ion in mg, obtained from the calibration curve for the corresponding sample absorbance.

(8) Expression of results on the basis of experiment

Results are expressed as mg/1 SO4.

(9) Precautions followed

(i) Time and rate of stirring should be kept constant throughout the determination i.e. for standards, blank and samples.

(ii) Final volume should be kept constant.

(iii) Make all turbidity measurements after a constant selected time, when the turbidity is maximum.

(iv) Wherever required use sample blanks to account for the interference due to turbidity and colour of the sample.

Suspended matter may be removed by filtration.

(v) Method holds good up to the concentration 60 mg/1. For higher concentration, dilute proportionately.

13. Estimation of Magnesium

(1) General Introduction

Magnesium is a common constituent of natural water. It is one of the important contributors to the hardness of water. Magnesium salts broke down on heating to form scale in boilers. Concentrations greater than 125 mg/1 can also exert a cathartic and diuretic action. The magnesium concentration in water may vary from zero to several hundred milligrams per litre, depending on the source of the water.

Chemical softening of ion exchange reduces the magnesium and associated hardness to acceptable levels.

(2) Estimation of Magnesium by calculation

For routine estimation of magnesium concentration in water, the magnesium is calculated from the EDTA.calcium and total hardness titration methods.

Magnesium = Total hardness - Calcium hardness (all expressed as mg/1 CaCO3)

14. Estimation of Calcium

(1) General introduction

The presence of calcium in surface water results from passage through or over deposit of lime stone, dolamite, gypsum and gypsiforous shale. The concentration of calcium may range from zero to several hundred miligram per litre depending on the source. Small concentrations of calcium carbonate prevents corrosion of metallic pipes by laying down a protective coating. On the other hand, appreciable quantity of calcium salts break down on heating, to form harmful scale in boilars, pipes and cooking utensils. Calcium contributes to the total hardness of water.

The atomic absorption spectrophotometer method is the most accurate method available for the calcium determination. But the EDTA and permagnate titrimetric methods also give good results for control and routine applications. Due to simplicity and rapidity of the EDTA method it has become first choice of the analyst. Hence, only EDTA titrimtric has been described.

(2) Principle observed

When EDTA is added to water containing both'calcium and magnesium it combines first with the calcium. By titrating with EDTA calcium can be determined directly, when the pH is sufficiently high sothat the magnesium is precipitated as the hydroxide and an indicator is used which combines only with the calcium. Indicator shows a colour change when all of the calcium is complexed by EDTA, at a pH of 12 to 13.

(3) Interference drown

Under conditions of test, the following concentrations of the ions causes no interference with the calcium hardness determination:

Copper 2 mg/1 Ferrous iron 20 mg/1 Ferric iron 20 mg/1 Manganese 10 mg/1

Zinc 5 mg/1 Lead 5 mg/1 Aluminium 5 mg/l Tin 5 mg/1

Orthophosphate will precipitate calcium at the pH of test. Stroncium and barium also interfere. Alkalinity in excess of 30 mg/1 may cause an indistinct and point with hard waters.

(4) Reagents required

Standard EDTA titrant, 0.01 M : Dissolve 3.723 g analytical reagent grade ethylenediminetetraacetic acid disodium salt. Na2 H2C10H12O8N2.2H2Oin distilled water, and dilute to 1 litre. Standarize against standard calcium solution. This solution is approximately 1.00 mg $CaCO_3$/ ml.

Standard Calcium Solution, 0.01 M : Weigh 1 g anhydrous calcium carbonate, CaCO3 power into a 500 ml erlenmeyer flask. Place a funnel on the flask and add slowly HC1 until all the CaCO3 has dissolved. Add 200 ml distilled water and boil for few minutes to expel out CO2, cool, add few drops of methyl red indicator, and adjust to the intermediate orange colour by adding 3N NH4OH or 1 + 1 HC1, as required. Make up the volume up to 1 litre in a volumetric flask. This standard solution is equivalent to 1 mg $CaCO_3$/1 ml.

(5) Indicator used

Mureoxide (ammonium purpurate) was the first indicator available for the detection of the calcium end point, but due to difficulty in recognizing the end point it is rarely used these days. Eriochrome Blue Black R or Solochrome Dark Blue (also called Calcon) gives an improve and point, changing colour from red to pure blue.

Eriochrome Blue Black R : Prepare a stable form of the indicator by grinding together in a mortar 200 mg powdered dye and 100 g solid sodium chloride to a 40 to 50 mash. Store in a tightly stoppared bottle. Use about 0.2 to 0.4 g of the mixture for each titration. During the course of titration the colour changes from red through purple to bluish purple tint. The pure blue with out any trace of reddish or purple tint. The pH of the some sample (not all) must be raised to 14 (rather than 12 to 13) by using 8N NaOH in order to get a sharper colour change.

Sodium hydroxide, NaOH, N : Dissolve 40 g sodium hydroxide in distilled water, and make up the volume to 1 litre. Store in polythene bottle.

(6) Procedure followed

Take 100 ml or smaller portion diluted to 100 ml so that calcium content is about 10 to 20 mg. Add 5 ml NaOH solution to produce a pH of 12 to 13. Stire and add about 0.2,0.4 g of indicator mixture. Because of the high pH required, titrate immediately after adding alkali and indicator. Add EDTA titrant slowly with continuous striking to the proper and point.

Carry out a reagent blank using 100 ml distilled water under identical conditions.

Standarize EDTA titrant with 25 ml of calcium standard. Follow the same method as described for the sample.

(7) Calculation derived

$$Calcium\ as\ mg/1\,CaCO_3 = \frac{(V_1 - V_2)\,X\,1000}{ml\ sample}\,X\,\frac{25}{V_3}$$

Where,

V1 is the volume of EDTA required for sample.

V2 is the volume of EDTA required for reagent blank.

V3 is the volume of EDTA required for 25 ml of standard calcium.

Expression of results : Results are expressed as mg/1 of CaCO3. Round off the results to the nearest whole number.

(8) Precision and Accuracy obtained

The precision and accuracy of about ± 2 to 4 % is expected from a good analyst, when the concentration is more than 25 mg/1. Precision and accuracy also depends on the quantity of calcium present.

(9) Precautions followed

(i) Titrate immediately after adding the indicator because it is unstable under alkaline conditions.

(ii) For reconising end point use a colour comparison blank prepared under identical conditions.

(iii) The final volume for sample and blank should be kept constant.

(iv) Use deionized distilled water for solution preparations, and any dilution necessary.

(v) Add same quantity of indicator to sample and blank, that the colour comparison is better.

15. Estimation of Sodium and Potassium

The Sodium and Potassium contents of sewage water are determined by flame photometer as suggested by APHA (1975).

Procedure adopted

10 ml of well mixed sewage water is taken and it is digested with cone, nitric acid till white residue on drying is left. After that, samples are dissolved in cone. HC1 and warm distilled water and after filtration the final volume is made upto 10 ml. This solution is atomized in flame photometer and the concentration of sodium and potassium are determined with the help of calibration curves and the results are expressed in ppm.

13

Technique Used For Study of Ecological and Vegetational Analysis

Method for Community Study

The study Plant Community is a part of the natural vegetation, it grows, develops and attend into a functional unit and is in a state of constant flux with the environment. It has trophic organisation and metabolic pattern.

An Ecological Community can be regarded as an assemblage of species populations that has the potential of interaction. More precisely, a community may be defined as an interactive assemblage of species occurring together within a particular geographical area, a set *e£* species whose ecological function and dynamics are in some way interdependent. These interaction include overt and critically determining compitative interaction and feeding relationships as well as more subtle manifestations, such as reliance of plant or animal for pollination and seed dispersal or of animals on plants for meeting habitat requirements.

Physiognomy : The form and structure for physiognomy of a Community can be described in a number of ways, each assessing somwhat different aspects, each having certain advantages and limitations, and as might be expected, each having its proponents in ecological circles.

Growth forms : Perhaps, the least technical characterization of a Community's structure particularly applied to plants.

Method of Dauserean (1951) Proposed –

(I) Six growth forms :
Trees, Shrubs, Herbs, Bryoids, Epiphytes, Lianas

(II) Three sizes :
Tall, Medium and Low.

(III) Four functions or leaf habits :
Deciduous, Semi-deciduous, Evergreen Succulent and TW Evergreen Leafless.

(IV) Six leaf shapes :
Needle or spine, Graminoids, Medium or Small, Broad, Compound, Thalloids.

(V) Four leaf texture :
Fibrary, Membranous, Sclerophyll, Succulent.

(VI) Four aspects of coverage :
Barren or very Sparse, Discontinuous, in Tufts or Groups, Continuous.

THE LIFE FORM

Avarians of plant growth form reffered to as life form, was developed at the turn of the twenteeth century by Danish botanist Raunkiaer (1934).

The categories *Raunkiaer* recognised based on the amount and kind of protection afforded to the buds and shoot apices which are as follows :

(I) *Phanerophytes :* The plants are wood, shrubs and herbs surviving buds or shoot apcies born on height.

(II) *Cheamephytes :* These are small plants, herbs and creeping woody plants.

(III) *Hemicryptophyte :* Herbs growing in Rosettes and tussocks.

(IV) *Geophytes :* Buds are burried in ground. Tuberous and bulbous herbs.

(V) *Hydrohalophytes* : Annual herbs.

THE STRATIFICATION

All the plants in the Community are not of the same size and do not occupy the same space there. The distribution of the plants in the vertical space is described as the stratification or layering.

Stratification can be studied in aerial and under ground regions.

THE ZONATION

Horizontal changes in the physical environment are reflected in zonational changes in plant and animal components of ecological communities. This is very useful in study of marine ecosystem, fresh water ecosystem and in relation to gradient.

THE HORIZONTAL DISPERSION

The horizontal spacing or dispersion of plants and/or animal can also be used to describe the structure of an ecological community. These are three basic patterns of dispersal :

(I) Random

(II) Uniform or Regular

(III) Clumped or Contagious

THE SOCIABILITY

The sociability express the relation of individuals to each other and indicates the closeness between individuals. Braun Blanquet (1932) recognised five arbitrary categories of sociability:

S1 = Plants growing singly

S2 = Plants growing in small groups

S3 = Plants occurring in small patches

S4 = Plants forming large patches

S5 = Plants occurring in essentially continuous populations.

Whitford (1949) has attempted to express sociability in quantitative terms by the following equations :

Sociability = Density x 100 / Frequency

AN VITALITY

It is concerned with the normal growth and reproductive ability of the species which helps it in maintaining its position in community. Various authors have recognised three to five categories of Vitality :

V1 = Plants which germinate but die soon without reproducing.

V2 = Plants which linger after germination but can not reproduce.

V3 = Plants reproducing but only veget'atively.

V4 = Plants reproducing sexually but rather feebly.

The Quantitative Characters :

(1) The Density

Density is the number of individuals of a species in an unit area or volume.

$$Density = \frac{Total\ number\ of\ plants\ in\ the\ quadrats}{Number\ of\ quadrats\ examined}$$

The Density is determined by actual count for each species in the community.

(2) Frequency

Frequency indicates the homogeniety of dispersion of the individuals of a species in the Community. The frequency is expressed as the percentage occurrence of individuals of a species in a number of observations.

$$\%\ Frequency = \frac{Number\ of\ quadrats\ in\ which\ the\ species\ occur}{Total\ number\ of\ quadrats\ examined} X\ 100$$

Raunkiaer (1934) divided the frequency into five classes :

A = 0 - 20 %

B = 21 - 40 %

C = 41 - 60 %

D = 61 - 80 %

E = 81 - 100 %

Raunkiaer proposed a law of frequency which states that the percentage of species failing under the five classes of frequency will be in the order

$$A > B > C > D > E$$

He found that in a homogenously distributed vegetation the proportion of the species will be 53, 14, 9, 8 and 16 per cent respectively in the five classes.

(3) Abundance

Abundance of a species in the Community represents its relative distribution in it. It is related to density but is a qualitative estimates.

$$Abundance = \frac{Total\ number\ of\ plants\ of\ a\ species\ in\ all\ the\ quadrats}{Number\ of\ quadrats\ in\ which\ the\ species\ occured}$$

(4) An importance Value Index (IVI)

Curtis and Mclhtosh (1950) proposed as Importance Value Index (IVI) which is sum of the relative values of the three quantitave characters. The relative value are expressed as percentage of the total values for all species in a community as given below :

(i) Relative Frequency (RF)

$$= \frac{Frequency\ of\ one\ species}{\sum Frequency\ of\ all\ species} X\,100$$

(ii) Relative Density (RD)

$$= \frac{Total\ number\ of\ plants\ of\ a\ species}{Total\ number\ of\ plants\ of\ all\ the\ species} X\,100$$

(iii) Relative Dominance (R. Dom.)

$$= \frac{Total\ canopy\ cover\ of\ species}{Total\ canopy\ cover\ of\ all\ the\ species} X\,100$$

(5)An index of Dominance

$$C = \Sigma\ (ni/N)^2$$

Where,

ni = Importance value of each species (number, biomass, production, etc)

N = Total of all Importance value.

(6) An index of Similarity between two Communities

$$IS = \frac{2C}{A+B}$$

Where,

A = Number of species in Community A

B = Number of species in Community B

C = Number of species common to both Communities.

IS = Index of Similarity

(7) The Point Centered Quarter Method of Sampling vegetation

This method has been developed by Cottan and Curtis (1949, 1956). In this experiment, four distances are measured at each sampling point. Four quarters are established at the sampling point through a cross formed by two lines. One line is the compass directed and the second line running perpendicular to the compass direction through the sampling point.

The plants nearest to the pin in the four quarters are recorded as shown in Fig. –

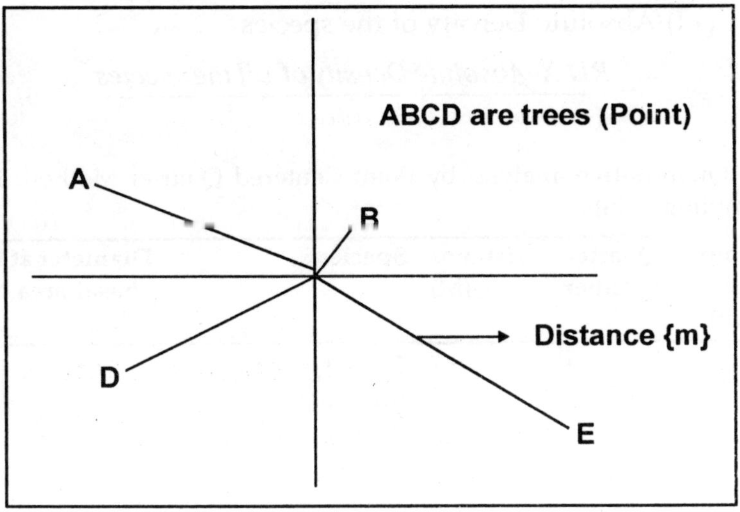

The important Community characteristics are computed as under :

(I) Frequency

$$= \frac{\textit{Number of points with species present}}{\textit{Total number of points observed}} X\ 100$$

(II) Average Distance

$$= \frac{\text{Distance of all species recorded}}{\text{Number of points X 4}}$$

(III) Average Area = (Average Distance)^2

(IV) Absolute Density of all the species

$$= \frac{100\,X\,100}{Average\ Area\,(Cm^2)}$$

(V) Relative Density of the species

$$= \frac{Number\ of\ individuals\ at\ all\ points\ X\,100}{Number\ of\ points\ X\ 4}$$

(VI) Absolute Density of the species

$$= \frac{RD\ X\ Absolute\ Density\ of\ all\ the\ species}{100}$$

An Quantitative analysis by Point Centered Quarter Method. Five Sampling Points

Sampling point	Quarter number	Distance (M)	Species	Diameter at the basal area cm^2
1	1	0.7	*Tectona grandis*	5.5
	2	1.6	*Acacia arabica*	42.5
	3	3.5	*Butea monosperma*	17.0
	4	2.0	*Diospvros melanoxylon*	25.0
2	1	1.1	*Tectona grandis*	4.0
	2	0.8	*Tectona grandis*	5.0
	3	1.9	*Tectona grandis*	5.0
	4	1.8	*Tectona grandis*	4.0
3	1	1.3	*Acacia arabica*	75.0
	2	0.7	*Tectona grandis*	3.0
	3	1.5	*Butea monosperma*	9.0
	4	2.0	*Butea monosperma*	23.0
4	1	3.1	*Acacia arabica*	14.0
	2	1.7	*Tectonagrandis*	6.0
	3	1.1	*Tectona grandis*	5.0
	4	1.9	*Acacia arabica*	12.0
5	1	2.5	*Acacia arabica*	23.0
	2	2.2	*Acacin arabica*	18.0
	3	1.4	*Tectonagrandis*	5.0
	4	2.8	*Butea monosperma*	25.0

Calculations derived

(A) Mean Distance (D) 35.6 / 20 = 1.78 m

(B) Absolute Density = Area / D^2

Where, D = Mean Distance

(C) Number of trees per 100 m² = 100 / (1.78)²
= 100/3.17
= 31.5

(D) An absolute Dominance
= means 'ba' per tree x number of trees in species
Where, ba = basal area

(E) Number of trees in species

	Species	Number of quater	Number of trees in 100 M²
1	Acacia *arabica*	6/20 = 0.3	0.3x31.5 = 9.5
2	*Butea monosperma*	4/20 = 0.2	0.2x31.5 – 6.3
3	*Diospyros melanoxylon*	1/20 = 0.05	0.05x31.5 = 1.6
4	*Tectona grandis*	9/20 = 0.45	0.45x31.5 = 14.2
			Total = 31.5

Mean basal area (ba) by species for the 20 lines :

(i) Acacia arabica

Diameter (cm)	Basal area (cm)²
42.5	1418
75.0	4418
14.0	154
12.0	113
23.0	415
18.0	254
Total basal area	6772
Mean basal area	1129

(ii) Butea monosperma

Diameter (cm)	Basal area (cm)²
17.0	227
9.0	64
23.0	415
25.0	491
Total basal area 1	1197
Mean basal area	299

(iii) Disopyras melanoxylon

Diameter (cm)	Basal area (cm)^2
25.0	491
Total basal area	491
Mean basal area	491

(iv) Tectona grands

Diameter (cm)	Basal area (cm)^2
5.5	24
4.0	13
5.0	20
5.0	20
4.0	13
3.0	7
6.0	28
5.0	20
5.0	20
Total basal area	165
Mean basal area	18

(F) Therefore Dominance of

	Dominance	Rank
Acacia arabica	1129 x 9.4 = 10613 cm2	1
Butea monosperma	299 x 6.3 = 1884 cm2	2
Diospyras melanoxylon	491 x 1.6 = 786 cm2	3
Tectona grandis	18 x 14.2 = 756 cm2	4

(G) Absolute Frequency

$$= \frac{Number\ of\ points\ with\ species}{Total\ points} X\,100$$

1	*Acacia arabica*	4/5 x 100 = 80 Per cent
2	*Butea monosperma*	3/5 x 100 = 60 Per cent
3	*Diospyras melanoxylon*	1/5 x 100 = 20 Per cent
4	*Tectona grandis*	5/5 x 100 = 100 Per cent
		260 Per cent

Determination of Importance Value Index

(I) Relative Density

$$= \frac{Number\ of\ individuals\ of\ species}{Total\ number\ of\ individuals} X100$$

(1)	*Acacia arabica*	9.5/31.5 x 100 = 30 Per cent
(2)	*Butea monosperma*	6.3/31.5 x 100 = 20 Per cent
(3)	*Diospyras melanoxylon*	1.6/31.5 x 100 = 5 Per cent
(4)	*Tectona grandis*	14.3/31.5 x 100 = 45 Per cent
		= 100 per cent

(II) Relative Dominance

$$= \frac{Do\min ance\ of\ a\ species}{Do\min ance\ of\ all\ species} X100$$

(a)	Acacia arabica	10,613 / 13,539 x 100 = 78.4 Per cent
(b)	Butea monosperma	1,884 / 13,539 x 100 = 13.0 Per cent
(c)	Diospyras melanoxylon	786 / 13,539 x 100 = 5.8 Per cent
(d)	Tectona grandis	256 / 13, 539 x 100 = 1.9 Per cent

(III)Relative Frequency

(a)	Acacia arabica	80 / 260 x 100 = 30.8 Per cent
(b)	Butea monosperma	60 / 260 x 100 = 23.1 Per cent
(c)	Diospyras melanoxylon	20/260 X 100 = 7.7 Per cent
(d)	Tectona grandis	100 / 260 x 100 = 38.5 Per cent

Summarized Table

(IV) Importance Value = RD + R.Dom. + RF

Species	RD	R.Dom	RF	IVI	IVI Rank
(a) Acacia arabica	30.0	78.4	30.8	139.2	1
(b) Butea monosperma	20.0	13.9	23.15	7.0	3
(c) Diospyras melanoxylon	5.0	5.8	7.7	18.5	4
(d) Tectona grandis	45.0	1.9	38.58	5.4	2

Conclusion direved

From above table indicates that Acacia *arabica* and *Tectona grandis* are the dominant and co-dominant. species in above Forest Community.

Practical Exercise Related to Biomass, A Production and Energy

(A) Biomass

= Total weight of OM (Above ground+below ground)) of particular area for given time.

$$\Delta B = B_2 - B_1$$

(B) Production rate

$$= \frac{W_2 - W_1}{t_2 - t_1}$$

Where,
W_2 is the dry weight at time t_2
W_1 is the dry weight at time t_1

(C) Turn Over rate

$$= \frac{Gross\ Primary\ Productivity\ (GPP)}{Standing\ Crop\ Biomass} hr/day/year$$

(D) Turn Over time

$$= \frac{Standing\ Crop\ Biomass}{GPP} hr/day/year$$

(E) Ecological turn over

$$= \frac{Community\ Respiration\ rate\ (R)}{Total\ Community\ Biomass\ (B)}$$

(F) Turn Over value (T)

$$= \frac{B_{max} - B_{min}}{Total\ B}$$

Where,
B_{max} is maximum biomass.
B_{min} is minimum biomass
Total B is total mean biomass

(G) % Efficency of Energy Utilization (EU)

$$EU = \frac{Gross\ Production}{Total\ Solar\ Radiatio} X 100$$

A Determination of Energy — Value of Materials by Bomb Calorimeter

The leith (1968) described the method of energy estimation with help of oxygen bomb calorimeter. The productivity values are then given in terms of energy. The weight measurements are not exactly comparable. Therefore, now a days, to give productivity values in terms of energy is, preferred than in terms of weight.

Method used : 1 g of powder of dried plant materials is taken in pellet machine. The pellet is now weighed alongwith ignition wire and is placed in ignition cup in the bomb. The known quantity of water is taken in Bomb Vessel and temperature is allow to constant by the help of stirrer. When temperature becomes constant, this will be t_1. The Calorimeter bomb filled with oxygen and placed inside the vessel before taking the reading of vessel. Now bomb is link with electrical circut for ignite the inside material. After ignition of material, there is rise in temperature of water. This is *tz*- The difference of the temperature is the rise of the temperature. The energy given is equal to energy taken.

MST = MST

Where,
M = Mass
S = Specific heat
T = Rise in temperature

(a) Energy (e)

$$= \frac{W1(t2-t1)+W2(t2-t1)}{W} Cal / g\ dry\ wt.$$

Where,

W1 = Water equivalant of the Cylinder (g)

W2 = Values of Water in the Cylender

W = Weight of pellet (Plant part or animal part)

T1 = Initial Temperature °C

T2 = Final Temperature °C

(b) Volume (v)

$$= \frac{W\,\Delta t - \sum c}{G} Cal / g\ dry\ wt.$$

Where,

G = Dry weight of material

At = Rise in temperature

W = Water equivalent of system

SC = Temperature rise value due to ignition.

A Reproductive Capacity Of Plants (Rc)

$$RC = \frac{Seed\ out\ of\ per\ plant\ X\ \%\ Germination}{100}$$

A Conservation Value Of Plant (Cv)

$$CV = 100\ (S_{wp}\ /\ S_{wo}\ x\ 100)$$

Where,

CV = Conservation value of species

Swp = Weight of soil (dry) washed from a plot under cover of species.

S_{wo} = Weight of soil (dry) washed from an identical but base plot under identical erosive factor.

Demographic Quotient (Cloud 1968, 1969,1970)

$$Q = \frac{Total\ Resource\ available}{Population\ Density\ X\ Per\ Capita\ Consumption}$$

Where,

Q = quality of life

Measurement of the Height of the Trees

For measuring the heights of standing trees a number of instruments have been devised, which are variously called dendrometer, hypsometer, altimeter, clinometer, etc. They all utilize either the geometrical properties of similar triangles or the trignometrical ratios of angle.

By Sextant we measured the Angle

$$Angle\ \theta = \frac{Perpendicular}{Base}$$

Base = Parpendicular x Angle θ (tan θ)

Parpendicular	=	Base	x	tan θ
(Height of the tree)		(Distance)		

A Measurement of Girth or Diameter of the A Tree

$$A = \pi r^2 = \pi d^2/4 = 0.7854\ d^2$$

Where,

A = is the Area

r = is the radius

d = is the diameter or Where,

or $A = g^2 / 4\pi = 0.07958\ g^2$

Where,

g = is the girth.

An Estimation of the Volume of Logs and A Trees

$$V = L X \frac{G}{4\pi} X \frac{4}{144} Cu. ft.$$

Where,

L = the length in feet of a log

G = the mid girth in inches

V = Volume

Complexity Index Determination of (Holdridge, 1967)

This is a very simple methametical expression for the quantitative characters of the 'Community in form of Complexity Index. This cofficient was used originally only for comparing the vegetation of larger areas in relation to rainfall.

CI = number of spp. x Density x Basal area x Height x 10^{-3} Based on 0.1 ha.

Where,

CI = Complexity Index

A Climate, Vegetation and Production Index A (CVP Index)

To measure the ability of a site for plant growth in a particular climate, and to express it as an Index and than to correlate it with site class. Peterson has given the formula

$$CVP\ Index\ or\ I = \frac{Tv}{Ta} X \frac{P}{12} X \frac{GE}{100}$$

Where,

TV = the temperature of the warmest months

T_a = the annual range of temperature of the warmest month

P = the annual precipitation in month

G =the length of growing season

E = the Evapo-transpiration reducer

Competition Principle

Gause (1934) who concluded from an experimental study on two species that two species with similar requirements can not co-exist and one of them is excluded.

The competition away the organisms may be for space, food, nutrients, water, light, gases, pollinating and dispersal agents or such other factors. Competition may be between the individuals of the same species (Intraspecific) or of different species (Interspecific).

$$\frac{dN_1}{dt} = r_1 N_1 \left(\frac{K_1 - N_1 - \alpha N_2}{K_1} \right) \text{ and}$$

$$\frac{dN_2}{dt} = r_2 N_2 \left(\frac{K_2 - N_2 - \beta N_1}{K_2} \right)$$

Where,

α = Competative Coefficient for species 2 in relation to species 1.

β = Competative Coefficient for species 1 in relation to species 2.

K = Carving Capacity of the environment

Stomatal Index Determination

$$SI = \frac{S}{S + E}$$

Where,

S = Number of stomata

E = Epidermal cells in a unit area

SI = Stomatal Index

Density of stomata may be calculated as number of stomata per unit area.

Study of Population Ecology

(a) Natality

It is the rate of births per unit time, The Natality is expressed by the formula

$$B = N_n/t$$ Where,

B = births

N_n = Number of new borns

t = Time

(b) Mortality

It is the death rate per unit time.

The mortality is expressed by the formula

$$d = D/t$$

Where,

d= death rate

D = Total number of deaths

t = Time

Glossary

Abaxial: The side or on the distant from the axis; as on the lower surface of a leaf.

Aberrant: A growth that deviates from the normal or usual type; as exceptional growths (aberrations) which occur in some tissue cultures.

Abort: To stop or fail to reach full development (maturity) or to develop imperfectly; as some hybrids are prone to embryo abortion or as some attempts to culture a plant may fail (abortive attempt).

Accession: Plant or seed sample, strain or population held in a genebank or breeding programme for conservation and use.

Acclimate or acclimatize: To adapt or adjust physiologically to a new environment or climate. The process is acclimation or .acclimatization and is regulated by natural processes or human cultural practices. Usage of acclimatize may be limited to adaptive responses involving changes in a selected environmental feature, such as light intensity.

Acetosyringone: One of the phenolic compounds released by wounded plant cells which can stimulate the transcription of the *vir* region of the *Ti* plasmid in the initial stages of transfer of DNA from *Agrobacterium* to the plant cell.

Adenosine Triphosphate (ATP): The general energy donor molecule in all organisms.

Adsorb: the adhesion of a liquid, gas or dissolved material to a solid surface, resulting in concentration of the adsorbate (the adsorbed material) to the adsorbent (the adsorbing agent, such as activated charcoal). The process is adsorption.

Adventitious: 1. Produced in an abnormal or unusual position, or at an unusual time of development. or away from the natural habitat; as when plant organs (buds, shoots, roots, others) develop on callus or nonzyotic embryos (embryoids) develop without an ovary or fertilization on callus (adventive embryony). 2. A manner of growth relying on adventive processes.

Adventitious: Developing from unusual points of origin, such as shoots or root tissues from callus or embryos from sources other than zygotes.

Aerate: To supply with or mix with air or gas. The process is aeration.

Aerobic or aerobiotic: Refers to an organism that lives in or a process occurring in the presence of molecular oxygen,

Agar or agar-agar: A gelatinous polysaccharide obtained from the red alga Gelidium corneum and from several other red algae. It is a solidifying agent that mixed with nutrient media (0.6-1%), forms a gel for growing tissue cultured plants and for other purposes. It ranges in quality from datively inert to very impure (complex, undefined). Its firmness as a gelling agent is affected by medium pH

(it is softer when the medium is more dilute). Agar gels melt at about 100°C but solidify at about 44°C.

Age: 1. The period in the life cycle of an organism; the process of growing older, mature. 2. The state of being old or senescent. 3. Culture age is a function of the number of subcultures and the time after subculture.

Agenesis: The absence of development.

Airlift fermenter: A cylindrical fermentation vessel in which the cells are mixed by air. which is introduced at the base of the vessel and rises through the column of culture medium. The cell suspension circulates around of the column as a consequence of the gradient of air bubbles in different parts of the reactor.

Algorithm: A description of how to solve a problem presented in terms that could be solved using a computer.

Aliquot or aliquot part: An evenly divided unit, portion or sample (fraction) of the whole.

Alkaline: A basic solution with a pH above 7.0.

Allele: An alternative form of a gene. Two or more genes are said to be allclcs when they: (1) occupy the same relative position (locus) on a pair of homologous chromosomes and undergo pairing during meiosis; (2) produce different effects on the same set of developmental processes; and (3) can mutate one to another.

Allopolyploid: hybrids between two (or more) species or genera containing the sum of the complete chromosome set of the different parental species, which can be a different ploidy (chromosome multiplicity) levels.

Allotetraploid: hybrids between two species or genera containing the sum of the complete diploid chromosome complements of each parent.

Alphanumeric: The set of symbols used to represent letters of the alphabet and the integers 0-9.

Amino acid: One of many organic acids containing a basic amino group (NH2), and an acidic carboxyl

group (COOH). Of several hundred naturally occurring amino acids only 20 are commonly found inprotein molecules, linked by peptide bonds. Some are commonly added to plant tissue culture media, especially glycine and glutamine.

Amphidiploid: hybrids between two species or genera containing the sum of the complete chromosome complements, which can be of different ploidy levels, of each parent.

Anther culture: Refers to culture of single pollen grains or of the anther, containing the male gametophytes or micropores, with the objective of producing monoploid plants.

Anthesis: The flowering period or efflorescence. This is the time of full bloom, which lasts till fruit set.

Anti-idiotype antibody: An antibody that has the properties of an antigen.

Antibiosis: The prevention of growth or development of an organism by a substance or another organism.

Antibiotic: One of many natural organic substances (or their synthetic analogues) secreted by plants or microorganisms that are toxic to other species, retard or prevent their growth and presumably function as defense mechanisms; as bacitracin, gentamycin, mycostatin, nystatin, penicillin, phosphomycin, rifampicin, streptomycin and terramycin etc. The phenomenon is antibiosis. Antibiotics are sometimes included in plant tissue culture media, with varying results, ranging from dramatic culture stimulation to induction of chromosomal instability.

Antibody: A protein (immunoglobulin) that is synthesized by a B lymphocyte and recognizes a specific sue on an antigen. The basic immunogbulin molecule consists of two identical heavy and two identical heavy and two identical light chains.

Antigen: A compound that induces the production of antibodies.

Antioxidant: A substance (such as ascorbic acid, citric acid or others) which is sometimes added to the sterilizing solution or to isolation medium to inhibit or prevent oxidative browning of the culture medium, due to bleeding of phenolic exudates. The latter may lead to tissue necrosis and death. Ascorbic acid (100 mg/1) and citric acid (150 mg/1) are most commonly used in sterilizing solutions.

Apex : The most extreme point of growth of a plant; as the apical shoot and root tips are located at the apexes (apices), and contain the apical meristem.

Batch culture : A cell suspension grown in liquid medium of a set volume. Inocula of successive subcultures are of similar size and cultures contain abou the same cell mass at the end of each passage. Cultures commonly exhibit five distinct phases per passage; a lag phase follow inoculation, then an exponential growth phase, a linear growth phase, next a deceleration phase and finally a stationary phase.

Bcteriostat: A chemical or other agent that does not kill but prevents bacterial growth and multiplication; is bacteriostatic.

Bergmann, L. (1960): He was first to obtain callus by plating cells from suspension cultures onto solid medium.

Bioassay: A biological assay performed on living cells or on a living organism; sometimes used to detect minute amounts of substances which influence or are essential to growth.

Biosynthesis: Biological synthesis, the building or forming of biochemical compounds in a living organism

Bleach: A fluid, powder or other whitening (bleaching) or cleaning agent. Bleach contains calcium hypochlorite or sodium hypchlorite and is a common disinfectant used for cleaning working surfaces, tools and plant materials in plant tissue culture.

Bridge: A filter paper or other substrate used as a wick and support structure for a plant tissuer culture when liquid media are used.

Buchner funnel: A porcelain funnel with a perforated flat circular base employed in vacuum filtration. After E. Buchner (1860-1917).

Bud-sport or bud mutation: A somatic mutation in a bud giving rise to a branch, fruit or flower that is a typical of the plant on which they occur. These characters (if beneficial) may be retained through vegetatrivc propagation from the affected area.

Bud: 1. An embryonic or undeveloped, enemerged stem, leaf or flower, often enclosed by reduced or Specialized leaves called bud scales. The beginning of incipient growth or development. 2. To graft onto another type of tree, plant. 3, A vegetative outgrowth from a yeast.

Buffer: A substance in solution, or system that resists pH change, (despite addition of some small amount of acid or base) and withstands shock; as buffered media can resist pH drift. The degree to which pH change is resisted is a measure of the solution's buffering capacity.

Bung: A stopper or closure.

Burgeon: To flourish, bud, sprout or grow.

Cabinet (growth cabinet): A cupboard suitable for incubating a small number of culture vessels or tubes under controlled environments. The degree of control over light, temperature or relative humidity is a function of quality of the cabinet but is generally less than that implied by the use of the term incubator or laboratory incubator.calcium alginate: A hydrophyllic colloid obtained from seaweed. A derivative of alginic acid ($C_6H_8O_6$)n. It reversibly immobilizes plant protoplasts in which condition they can tolerate a reduced osmotic potential.

Calcium hypochlorite ($Ca(OCl)_2$-$4H_2O$ mw 215): An algicide, bactericide and fungicide capable of disinfecting and bleaching. It is used in dilute solution (5-10% w/v) as a plant tissue disinfectant often when agitating or using a vacuum. Tissue damage occur if contact is prolonged. Thorough washing with sterile water generally follows treatment.

Calcium pantothenate or pantothenic acid (Ca(C9H16NO5)2 mw 476.54): The calcium salt of vitamin B_5. An occasional vitamin additive to plant tissue culture media. Solutions of this vitamin are not stable to autoclaving.

Callus culture: The cultivation of callus, usually on soldified medium and initiated by inoculation of small explants or sections from established organ or other cultures (inocula). Callus may be maintained indefinitely by regular subdivision and subculture. It may be used as the basis for organogenetic (shoot, root) cultures, cell cultures or proliferation of embryoids.

Callus, pi. *calli* or calluses: 1. Wound tissue, tissue formed on or below a wounded surface. 2. Disorganized tumor-like masses of plant cells that form in culture. These proliferate in an irregular tissue mass, and vary widely in texture, appearance and rate of growth; a function of the tissue type (species and explant) and the composition of the medium, the process of callus formation is callogenesis.

Cannula: A small tube for insertion into tissue; as a cork borer.

Carbon source: A source of the non-metallic element carbon (C); as organic substances like sugars taken up and metabolized by plant tissue cultures.

Casamino acid: An amino acid obtained through the digestion (acid hydrolysis) of the milk protein casein. Protein digest yields casamino acids and other substances. These are undefined constituents of some plant tissue culture media.

Casein: the principal protein of milk, a phosphoprotein.

Caulogenesis: Shoot formation; as de novo shoot development from callus.

Cell count: The number of cells per unit suspension volume or callus weight. Tissue is treated with chromic acid (5-8%) or pectinase (0.25%) for up to 15 minutes followed by mechanical dispersion, then cell numbers are estimated with a hemocytometer (haemocytometer).

Cell culture: The culture of single or groups of cells on solid or dispersed in liquid nutrient media (cell suspension cultures) usually following a set of defined growing conditions (protocols).

Cell hybridization: Formation of synkaryons, viable cell hybrids produced through cell fusion, they are identifiable by their increased chromosome number compared to the parent cells and the possession of characters found in one or another of the parental cells.

Cell line: Developmental history or descent, through cell division from a single original cell; as callus may constitute a group of cell, all descendants from a single cell plating. Numerous cell lines may be present in a culture. Any deviation in culture technique may favor one cell line over another.

Cell number: The absolute number or approximation of the number of cells per unit area of a culture or medium volume.

Cell selection: Selection within a group of genetically different cells; involves competition between cells often under some stress. Selection criteria may involve cell viability 1 a unique phenotype or biochemical activity; or another basis for choice. Select cells or cell lines are generally relocated to fresh medium for continued selection and often are exposed to an increased level of the stress agent. The final objective is usually to regenerate plants from those select cells, in hope that the plants will exhibit the traits selected for at the cellular level.

Cell suspension: Cells and small aggregates of cells suspended in a liquid medium; as in cell suspension cultures. Explants, or callus derived from them are transferred to liquid medium and the cultures are then agitated on a mechanical shaker. The ensuing single cells and small cell clusters are used for a number of purposes in plant tissue culture; as in single cell cloning.

Chromic acid or chromium trioxide (003 mw 100.01): This compound exists only in solution and is made by mixing potassium dichromate and concentrated sulphuric acid. A dilute solution may be used to prepare tissues for hemocytometer (haemocytometer) counts to determine cell numbers in cell suspension cultures. Concentrated solutions were once used for cleaning new or dirty glassware but have now been supplanted by detergents.

Clonal multilication or clonal propagation: Vegetative (asexual) propagation from a single cell or plant. See clone.

Clone: 1. A basic category of cultivar. The original cultivar or variety. Not to be confused with source clone, indicating different originals within the clone. Members have a common origin, are the extension of a single cell (via mitosis) or plant and have been produced by vegetative means only. Genetic uniformity is accepted. A clone may have one or more sources. 2. While the expectation is that members of a clone are both phenotypically and genotypically identical, this assumption is not always true and members may not be genetically or phenotypically equivalent (varients). the objective of much tissue culture propagation is to propagate very large numbers of selected plants with the same genotype. The process is cloning. The prefix men reflects the explant used to initiate the clone (meristem tip source clone). In a calliclone cloning involves the callus stage. 3. Specific DNA sequences are said to be cloned when isolated and propagated.

Closed continuous culture: A cell suspension culture with a continuous influx of fresh medium, maintained at constant volume by the efflux of spent medium. All cells are retained within the unit. See continuous culture.

Co-culture: The joint culture of two or more types of cells; as a plant cell and a microorganism or two types of plant cells; as is done in various dual culture systems or the nurse culture technique.

Another plant trait or traits not expressed in culture.

Crown gall: A disease of plants in which tumors form. The causal agent is the bacterium. *Agrobacterium tumefaciens.* Gall tissue has been grown in vitro. A tumor inducing portion of the bacterial genome (the Ti plasmid) my be used experimentally as a genetic tool to incorporate (vector) genetic information into plant cells.

Culture alteration: A persistent change in a cultured cell or tissue anatomy or physiology, including a change in one or more nutritional requirements or a change in its proliferative capacity.

Culture medium: A prepared nutrient solution (substrate) that may be chemically defined for growing plant tissues or other organisms in vitro.

Culture room: A controlled-environment room (light, temperature, relative humidity, often air conditioned, etc.) for incubation of plant cultures.

Culture tube: A test tube used to contain medium and an explant or culture derived from it, usually ' incubated in a controlled environment.

Culture vessel: A container used to hold medium and an explant or culture derived from it, usually incubated in a controlled environment.

De novo: (Latin: *fromthe beginning, anew).* Arising, sometimes spontaneously, from unknown or very simple precursors.

Decereration phase: the declining growth rate phase following the linear phase and preceeding the stationary phase in most batch suspention cultures.

Decontaminate: To free from contamination or surface sterilize.

Dedifferentiation: The resumption of meristematic activity by more or less mature cells through a reversal of the process of cell or tissue differentiation. Cell division leading to the formation of small, microvacuolate, isodiametric cells with prominent nuclei, that are capable of organized development; as when adventitious buds or roots develop on mature tissue.

Deficiency: 1. An insufficient supply or unusable form, of one or more nutritional or environmental requirements, for plant development, growth or physiological function. The absence of adequate conditions for plant growth or performance, resulting in disease. Z.The deletion of a gene or a series of genes. This may result in a mutant phenotype or may be lethal.

Dehumidifier: An apparatus that removes moisture from the air.

Deionized water: Water that has been passed through an ion exchange device to remove soluble minerals and some organic salts. The process is deionization.

Deletion: 1. Omission, removal or cancellation of some factor. 2. The lost portion of a chromosome or a nucleotide sequence in nucleic acid.

Derepression: The mechanism by which repression is alleviated; as when a repressing metabolite is removed, resulting in the increased level of a protein or enzyme.

Detergent: A cleaning agent; one of numerous synthetic cleaning preparations chemically different from soap. Detergents are commonly used to prepare work areas for aseptic plant tissue culture manipulations and to remove dirt and microorganisms from plant material prior to explantation.

Determination or topophysis: 1. The process of commitment to a specific pathway of development (cell, tissue, organism, growth form). This commitment may occur well in advance of the appearance of the phenotype.The perpetuaton of the phenotype provides an example of the stability of this process. 2. A phenomena in which explanted meristems and ensuing growth derived from them in culture, taken fromdifferent areas of a plant representing different phases, perpetuate the different phase phenotypes of that plant; as meristems from thorny (juvenile) citrus give rise to thorny individuals unlike meristems explanted from adult tissue that perpetuate the thomless condition in

Direct embryogenesis: Embryoid formation directly on the surface of zygotic or somatic embryos or on seeding plant tissues in culture, without an intervening callus phase.

Direct organogenesis: Organ formation directly on the surface of relatively large intact explants, without an intervening callus phase.

Dispense: To give, deal or porton out; as nutrient medium is portioned into glassware for plant tissue culture. Dissect: To cut, expose, separate or divide the parts of a plant or animal into sections.

Embryo culture : 1. Denotes a culture in which the explant was n embryo. Embryo cultures have been used to obtain viable offspring from seeds with a tendency for embryo abortion or when viable seed are limited in number. 2. Cultures in which embryos are induced to form (embryogenesis) whether in suspension or on a variety of explants or cultures on solidified media. More correctly, these nonzygotic or somatic embryos are termed embryoids.

Enation: Outgrowth on a plant surface.

Environmental chamber: A controlled environment cabinet (incubator) in which temperature, light quality, intensity and duration; and preferably also the relative humidity and airflow are controlled.

Enzyme: Any one of a number of spoecilized proteins produced in living cells, that speed the rate of specific chemical reactions, even at very low concentrations (organic catalyst), but is not used up in the reaction.

Equimolar: The same amount of solute per litre of solutiibn, the same molar concentration.

Established culture : 1. Achievement of an aseptic viable explant, synonymous with Stage I culture (culture establishment) or transplant, synonymous with Stage IV culture (establishment in soil). 2. A suspension culture subjected to several passages, adjusted so that cell number per unit time is consstant from subculture to subculture.

Ex vitro: (Latin;/row *glass).* Organisms removed from culture and transplanted, generally to soil or potting mixture.

Excise: To cut, sever or otherwise remove or extract an organ or a segment of tissue from a plant or plant part; as the surgical removal of shoot tips. The process is excision.

Explant donor: The source plant or mother plant from which the explant used to initiate a culture is taken.

Explant: The excised plant portion used to initiate a tissue culture. The process of dissection and removal to culture of these small organs or tissue sections is explantation. Explant choice, the timing of excision and pretreatment are important determinants of culture success.

Exponential phase: A phase in culture in which cells undergo their maximum rate of cell division. It follows the lag phase and preceeds the linear growth phase in most batch propagated suspesion cultures

Exude: To discharge slowly, leak liquid material (exudate) sometimes through pores or cuts but often by diffusion into the medium. This process of exudationis associated with a lethal browning of explants in some woody plant species. These in vitro exudateshave not been chemically characterized but are often referred to as tenins or oxidized polyphenols.

Factor: 1. A condition, influence or constituent to be taken into account. 2. A genetic character determinant (gene).

Field test: An evaluative test whereby the field performance of experimental plants is assessed in comparison to controls.

Filter sterilize: to sterilize by passing a solution through a porous material capable of separating out suspended microbes or their spores; as heat-labile components of nutrient media are sterilized.

Fixative: A compond that stabilizes, sets or fixes other compounds or structures securely so that their structural integrity is retained; as the process of fixation utilizes chemical agents to permanently prepare cells or tissues for microscopy.

Flame sterlize: To sterilize tools or instruments usually by heating them in a flame until they glow. This procedure is used in conjunction with ethanol (70%) emersion or ethanol (95%) dip. The most common means of sterilizing forceps and scalpels used in sterile tissue culture manipulations.

Floccule: An aggregation (coalescence) of microorganisms or colloidal particles floating in or on a liquid. Flocculation is seen in some contaminated liquid media appearing as a cloud.

Fluid drilling: A mechanical procedure for planting seed; pre-germinated seeds are suspended in a gel and

sowed through a fluid drill seeder. This technology is potentially adaptive for sowing artificial seed (somatic embryos or embryoids).

Genetic selection: Selection of genes, cells (cell selection), clones, etc., by man within populations or between populations or species. The usual purpose is to alter a specific phenotypic character. Such selection usually results in differential success rates of the various genotypes, reflecting many variables including selection pressure and genetic variability in populations.

Genetic transformation: The transfer of extracellular DNA (genetic information) among and between species; as, for example, with the use of bacterial or viral vectors.

Genetic variation: Differences in individuals derived from the same genotype in distincton to differences caused by the environment. Genetic variance denotes the proportion of pohenotypic variance caused by differences in the genetic make-up of an individual.

Genome: The basic haploid chromosome set of an individual; the sum of its genes. Genotoxic: Carcinogenic; toxic to the chromosomes. Geotaxis: Plant'orientation with respect to gravity

Germplasm or germ plasm: 1. The reproductive body tissues distinct from somatic (nonreproductive tissues). The genetic material, basis of heredity of an organism, passed on through previous generations. 2. An individual representing a type of species or culture that may bve held in a repository' for agronomic, historic (or other) reasons.

Gibberellic acid (gibberellin A3 or GA3, C19H22O6 nw 346.37): One of the gibberellins, a group of

growth hormones promoting cell division and elongation. The first of the group to be isolated and the most widely used gibberellin in plant tissue culture. Isolated from the fungal pathogen *gibberella fujikuroi.* It dissolves in base (ca. 1MKOH or NaOH-

Haberlandt, G.: An inspirational pioneer botanist who was the first to attempt to culture plant cells at the turn of the 20th century. He predicted the existence of hormones that would stimulate cell division (auxins) and suggested the possibility of exploiting the totipotentiality of plant cells.

Hanging droplet technique: *See microdroplet array technique.*

Hemicellulase: An enzyme from *Aspergillus niger,* available as a commercial preparation, that degrades hemicellulose to galactose.

HEPA filter: An acronym for high efficiency particulate air filter. A filter capable of screening out particles larger than 0.3 m. They are used in laminar air flow cabinets (hoods) for sterile transferr work.

Heterograft or xerograft: An interspecific graft. The likelihood of success of this graft is proportional to the degree of relatedness between the donor and recipient.

Heteroplasmon or heteroplast: A cell with a mixture of two types of cytoplasm (cytoplasmic hybrid); cell with foreign organells.

Hexitol: A sic carbon sugar alcohol, such as (meso-, myo-, or i-) inositol.

Hood: Laminar air flow cabinet.

Hydrate: To add or incorporate water. The process is hydration.

Hypertrophy: An abnormal increase in cell size causing irregular swelling oir growth. Hypertrophy is usually a disease or other stress-induced response.

Hypochlorite: A salt of hypochlorous acid (sodium hypochlorite, potassium hypochlorite or calcium hypochlorite). All are oxidizing agents used for disinfecting and for bleaching.

In vitro: (Latin = in *glass).* Experimentation on organisms or portions thereof in glassware or culture; growing under artificial conditions as in tissue culture. The antonym of in vivo.

In vivo: (Latin = in *life)* Experimentation on organisms under natural conditions within intact living organisms, the antonym of in vitro,

Incubate: To maintain under conditions favorable for development, often in an incubator. The process is incubation.

Incubator: An apparatus providing controlled environmental conditions, (light, photoperiod, temperature, humidity, etc.) suitable for (incubating) plants or plant cultures. This term is sometimes used as as synonym for cabinet (growth cabinet) but usually implies a greater degree of environmental control.

Indirect embryogenesis: Embryoid formation on callus tissues derived from zygotic or somatic emtryos. . seedling plant or other tissues in culture. The antonym is direct embryogenesis.

Indirect organogenesis: Organ formation on callus tissues derived from explants. the antonym is direct organogenesis.

Infestation: Occupation by threatening numbers of insects, mites or potential disease agents; as mite infestations in incubators. An infesting organism can cause contamination of tissue cultures. This should not be confused with infected, as infection is not necessarily implied.

Infrared gas analyzer (IRGA): An instrument for measuring the proportion of a particular gas in a mixture: as changes in *CO2* concentration can be monitored to evaluate a plant's photosynthetic or respiratory activity.

Innoculum: 1. Material introduced (inoculated) onto or into a host or medium. 2. Potential inculum implies that it is infective and may by chance result in natural inoculation of a host.

Inoculate: 1. To deliberately introduce something into; as inoculum is placed into (or onto) medium to initiate a culture. This process is inoculation, but should not be confused with contaminate 2. Vaccinate.

Jiffy pots: A brand name for peat pots that are sometimes used for ex vitro transplantation.

Juvenile: The immature, usually non-reproductive (reproductively incompetent), sometimes phenotypically distinct phase of plant growth. Juvenility should not be confused with young in age although young plants may also be in a juvenile phase. Juvenile phase tissue has repeatedly been foiund to perform better in culture than adult phase tissue. It is more likely to grow and to form callus, embryoids, shoots etc.

Lag phase: 1. Used to describe the first of five growth phases of most batch propagated cell suspension cultures in which inoculated cells in fresh medium adapt to the new environment and prepare to divide.

Laminar air flow cabinet or hood: A structure in which a uniform flow of filtered air prevents settlina of particles in the work area. The air is filtered through prefilter (furnace-filter quality), then a high efficiency paniculate air (HEPA) filter that strains out particles greater than 0.3 m. These hoods are commonly employed for aseptic plant tissue culture manipulations and are designed to accommodate one to the several persons, depending on the model.

Layering: 1. Covering stems, runners or stolons with soil causing adventitious roots to form at the nodes enabling propagation by rooted cuttings. This procedure is used commercially to propagate many plants, such as the brambls. 2. In vitro layering involves the horizontal placement on agar of cultured shoots (with or without leaves) or nodal segments to promote axillary bud proliferation.

Lecithin: A naturally occurring, choline-containing phospholipid present in animal and plant tissue. Chemically lecithins are similar to fat, but they also contain phosphorus and nitrogen.

Limiting requirement or factor: An environmental variable whose absolute levels at given time limits the growth or other activity of an organism.

Linear phase: The constant increase in cell number following the exponential growth phase and proceeding the deceleration phases in most batch suspension cultures.

Liquid culture: The culture of plant cells on liquid medium, in suspension or on supports. Cultures are either held steady (stationary culture) or are shaken (agitated culture or shake culture).

Liquid medium: Medium not solidified with a gelling agent. Liquid media are used for suspension cultures and for a wide range of research purposes. They are also useful for Stage I culture in some micropropagation protocols. Usually a support structure or wick is used to hold the tissue above the nutrient medium.

Maternal inheritance: Inheritance controlled by extrachromosoml (cytoplasmic) hereditary determinants. MDA: microdroplet array technique.

Medium formulation: One of many available, usually derived formulas that are used in plant tissue culture. The formulas commonly contain the microelements and microelements (high and low salt formulas are available), some vitamins (B vitamins, inositol), hormones (auxin, cytokinin and sometimes gibberellin), a carbohydrate source (usually sucrose or glucose) and often other substances such as amino acids (the most common addition is glycine) or complex growth factors. Media may be liquid or solidified with agar, the pH is adjusted (ca. 5-6) and the solution is sterilized (usually by filtration or autoclaving). Some formulations are very specific in the kind of explant or plant species that can be maintained some are very general.

Medium: The substrate for plant growth; as nutrient solution, soil, sand, etc. This is a general term for the liquid or solidified formulation upon which plant cells tissues or organs develop in plant tissue culture.

Mericlinal: A meristem tip source clone. This term can be synonymous with calli clone or may not involve callus.

Meristem culture: The culture of meristems; meristemaitic dome tissue without adjacent leaf primordia or stem tissue. In practice this usually means meristem tip culture. This may more generally imply the culture of meristemoidal regions of plants or of meristematic growth (associated with or sharing the characteristics of the meristem) in culture.

Meristem tip culture: Cultures derived from meristem tip explants. They are excised for virus elimination or axillary shoot proliferation purposes, less commonly for callus production.

Meristem tip: A common explant comprised of the meristem (meristematic dome) and usually one pair of leaf primordia. Also refers to apical (apical meristem tip) or lateral (lateral or axillary meristem tip) origin. The term meristem tip is often confused with the term shoot tip, which is much larger and usually has more immature leaves and some stem tissue.

Meristem: A localized region of continuing mitotic cell division (meristematic cells), of protoplasmic synthesis and tissue initiation. From these undifferentiated tissue new cells arise that differentiate into specialized tissues. Meristems are located at the apical (shoot and root tips), axillary, marginal or lateral (cambia) and other growing points. In addition, in vitro meristems may occur within more or less differentiated callus tissue and are termed meristemoids. The meristematic dome without any leaf primordial tissue is sometimes used as an explant in virus elimination work but usually a meristem tip (meristematic dome plus one pair of leaf primordia) is the explant.

Meristemming: the utilization of or process of utilizing meristem tips for explants. This term is often used to refer to micropropagation via axillary shoot proliferation.

Meristemoid: A cluster of small, isodiametric meristematic cells with a meristem, or cultured tissue, with the potentialk for developmental (totipotential) growth. These cells have a dense microvacuolated cytoplasm and a high nucleo-cytoplasmic ratio. They may give rise to plant organs (shoots, roots) or entire plantlets in culture.

Micrograft: An in vitro graft. This procedure involves placing a meristem tip or shoot tip explant onto a decapitated rootstock, that has been grown aseptically from seed or micropropaged. The purpose may vary, from virus elimination, using meristem tips and virus-free rootstocks, to viral assays or to erxpedite grafting that is usually done in the greenhouse. The explanted scion is placed directly on the cambium of the severed stock or inserted into a T or inverted T-shaped epidermal incision onthe stock. This technique has been used to eliminate viruses, viroids and mycoplasmas from Citrus species.

Micropropagation: Refers to propagation in culture by axillary or adventitious means. It is a general term for vegetative (asexual) in vitro propagation. It sometimes refers specifically to axillary bud proliferation.

Microwave oven: An oven in which heating results from microwave radiation of the food or contents. Such ovens are commonly employed to melt agar in making up plant nutrient media and may be useful to sterilize liquids and some plastic or glass items.

Minimum effective cell density: The inoculum density below which the culture fails to give reproducible cell growth. The minimum density is a function of the tissue (species, explant, cell line) and the culture phase of the inoculum suspension. Minimum density decreases inversely to the aggregate size and division rate of the stock culture. It is important to know this ratio when plating suspension cultures for selection of single cell lines so that colonies will not overgrow one another prior to obtaining readily manipulable size.

Mitotic index (MI): The ratio of nuclei undergoing mitosis (including prophase) to total nuclei. Ml = (Number of nuclei in mitosis/Total number of nuclei examined in the sample) X 100. A Ml of 03 means that 30% of cells in the population are observed in mitosis.

Mitotic nondisjunction: Occurs when sister chromatids fail to migrate to opposite poles of the cell during mitotic anaphase. The result is daughter cells with hyperploid and hypoploid chromosome counu.

Mixoploid: Refers to cells with variable (euploid, aneuploid) chromosome numbers; as mosaic or chimaenl components that differ in chromosome number, or the result of a variety of mitotic irregularities.

Mold: Any fungus growth; as on plants or in cultures.

Monoculture: A one-crop agriculture system. An intensified monoculture couJd refer to a cloaal (one genotype) agricultural system.

Mutability: the propensity of an individual's genes or genotype to undergo heritable mutation.

Mutagen: A chemical or physical treatment or agent capable of producing genetic mutation. Causes insertion, deletion or alteration of a base or part of the nucleic acid chain (DNA). Artificial mutations can be induced in cells using chemicals or ionizing radiation. The process is mutagenesis. Electromagnetic and particle radiation, such as UV, X-rays and b particles all increase mutation frequency. So do chemical mutagens such as ethylmethanesulphonate, acridine, nitrous oxide and many others

Natural complex: A complicated, non-synthetic, often undefined addendam in plant tissue culture media, such as orange or tomato juice or coconut milk.

Neoplasm: Localized cell multiplication or tumor; a collection of cells which have undergone genetic transformation. These cells differ in structure and function from the original cell type.

Neoteny: The occurrence (retention) of juvenile characters in the adult state or vice versa.

Newcommer's fluid: A fixative used in chromosome analysis composed of 6 parts isopropyl: 3 parts propionic acid: 1 part petroleum ether:! part acetone: 1 part dioxane (with or without ferric acetatel.

Nucellar embryony: The process by which individual cells of the nucellus give rise to somatic embryos (embryoids).

Plants on which this occurs naturally are considered good candidates for tissue culture of somatic embryoids.

Nucellus; An enzyme that can hydrolyze intemucleotide linkages of a nucleic acid.

Nucleo-cytoplasmic ratio: The ratio of cell nuclear to cytoplasmic volume. This ratio is elevated in meristematic cells and low in differentiated cells.

Nurse: A culture technique or the callus upon which a filter paper is placed separating single cells from the callus in the paper raft technique of Muir (1953). The callus (nurse tissue) releases growth factors and nutrients that induce growth in a single cells supported by the filter paper (those being nursed) and sharing the communal environment.

Nutrient film technique (NFT): Hydroponic plant growth whereby plant roots are suspended in shallow, slowly circulating nutrient solutions delivered as a continuous film of liquid rather than in bursts of liquid.

Nutrient gradient: A diffusion gradient of nutrients and gases is set up in tissues where only a portion of the tissue is in contact with the medium. This is common in callus culture and causes differential growth rates and senescent regions of callus. Similarly, diffusion gradients of nutrients are set up in the medium adjacent to the tissue. Gradients are less likely to form in liquid media.

Ontogeny: The developmental (ontogenetic) life history of an organism, its development from fertilized egg (zygote) to seed-forming adult (reproductive phase), with emphasis on embryonic development.

Open continuous culture: A cell suspension culture with a continuous influx of fresh medium, maintained at constant volume by the efflux of cells and spent medium.

Organ culture: The growth in aseptic culture of plant organs such as roots or shoots, beginning with organ primordi or segments and maintaining the characteristics of the organ.

Organic cosolvent: Compound used to dissolve some neutral organic substance; as in media preparation. These include alcohol (usually ethanol), acetone and dimethylsulfoxide (DMSO).

Organized growth: The in vitro development of organized explants such as meristem tips or shoot tips, floral buds or organ primordia, or their de novo formation from unorganized tissues,

Organogenesis: The initiation (de novo) and growth of organs (roots and shoots, usually) from cells or tissues; as in organ culture. Organs may form on the surface of explants (direct organogenesis) or upon an intervening callus phase (indirect organogenesis).

Ovulary culture: A culture in which the explant is an ovary containing the ovule(s). The ovules may be fertilized in culture (in vitro fertilization). This technique is used primarily when the ovary is essential to proper embryo development.

Ovule culture: A culture derived from an explained ovule: which may be fertilized in culture (in vitro fertilization). This technique is used to study development of zygotes and young embryos and is sometimes used to rescue embryos susceptible to abortion when embryo culture is not possible.

Packed cell volume (PCV): A quantitative method of estimating cell growth. It is based on the total cell volume in an aliquot of suspension culture. The aliquot is centrifuged for 5 minutes at *200* g after which the packed cell volume is expressed as a percentage of the aliquot volume.

Panicle culture: Aseptic culture of grain panicle segments, usually in an effort to induce microspore development.

Papain: A water soluble proteolytic enzyme (protease) extracted from papaya fruit and used especially as a meat tenderizer.

Paper raft technique: A technique developed by W.H. Muir (1953) to promote development of single cells taken from suspension cultures in which cells are placed onto filter paper squares set on actively growing callus (nurse tissue). Growth factors and nutrients from the callus tissue diffuse through the filter paper, promoting cell growth and development.

Particle radiation: Refers to - particles (positively charged) and B-particles (negatively charged), electrons. protons and neutrons. These particles are used to produce mutant cells or organisms in plant tissue culture.

Parts per million(ppm): This term has largely been replaced by the equivalent mg/1 of solution or mul/1 for liquids or gasses.

Passage number: The number of subculture intervals. Culture age is a function of passage number and the culture dilution ratio.

Passage time: The interval between subcultures or the culture period.

Pasteur pipette: A causal agent of a disease; as a pathogenic fungus, bacterium or virus.

PCV: packed cell volume.

pH drift: A shift in pH that occurs as cultures grow and is related to the buffering capacity of the medium.

pH: A measurement of the degree of acidity or alkalinity of a solution. The negative log of the hydrogen ion concentration, in moles perr litre (M/l). On the pH scale from 0-14, there is a 10 fold difference for each unit of change of pH. The lower the value, the more alkaline (more hydroxyl ions it contains). Most culture media are adjusted in the range pH 4-6 using 1M NaOH or HC1.

Phenocopy: A non-hereditary phenotypic change that is environmentally induced, during a limited developmental phase of an organism, that mimics the effect of a known genetic mutation.

Phenolic oxidation: The process by which many plant species which contain phenolics blacken through oxidatin when they are wounded; as in vitro, soon after explantation or subculture. This may lead to growth inhibition or, in severe cases, to tissue necrosis and death. Antioxidants are incorporated into the sterilizing solution or isolation medium to prevent or reduce oxidative browning.

Phenols or phenolics: A class of organic compounds with hydroxyl group(s) attached to the benzene nng forming esters, ethers and salts. Phenolic substances may bleed from newly explained tissues, oxidizing to form colored compounds visible in nutrient media or on the filter wicks supporting the tissue

Plasmotype: A cell type displaying features which are expressions of cytoplasmic, rather than nuclear inheritance.

Plasticity: The range of environmentally-inducible phenotypic expressions.

Plating efficiency: An estimate of the percentage of viable cell colonies developing on an agar plate relative to the total number of cells spread onto the plate. Plating efficiency is a function of the tissue (species, explant and cell line); medium composition; plating density and the phase of the stock culture.

Pleiotropy: The condition in which several characteristics are effected by a single gene.

Pollen culture: The culture of pollen grains, which germinate in vitro. Such cultures may eventually form monoploid callus, from which shoots or embryoids develop into monoploid plants.

Protoclone: Distinct phcnotypic regenerates from a plant protoplast. A clone initiated from a protoplast or protoplast-fusion product.

Protocol: A sequence of activities, techniques or procedures linked to achieve a specific goal; as a schedule for protoplast preparation.

Protoplast culture: The isolation and culture of plant protoplasts by mechanical means or by enzymatic digestion of plant tissues or organs, or cultures derived from these. Protoplasts are utilized for selectio or hybridization at the cellular level and for a variety of other purposes.

Protoplast fusion: The coalescence of the plasmalemma and cytoplasm of two or more protoplasts in contact with one another. Initial adhesion is a random process but coalescence may be promoted in various ways (induced fusion). When adhesion occurs between adjacent protoplasts during enzymatic wall degradation or between freshly isolated protoplasts in the absence of a fusion agent, it is termed spontaneous fusion.

Pure culture: Axenic culture.

Pure line: All cell or individual members are homozygous for one or more characters or genes and will give rise to more cells or organisms with the character(s) under consideration.

Radiation: Rays of heat, light or particles in wave form.

Ramet: The vegetatively propagated offspring of an ortet. An individual of a clone.

Reaction norm: The range of phenotypic responses of a particular genotype in response to the environmental influences.

Reciprocating shaker: A.plateform shaker with a back and forth action used for agitating culture flasks as variable speeds.

Reconstructed cell or recon: A viable cell hybrid, cybrid or a transformed cell resulting from geneuc engineering.

Reculture or subculture: The aseptic transfer of a culture or a portion (division when necessary) of that culture (inoculum) to fresh nutrient medium.

Reinert, J. (1958): Among the first to observe adventative embryogenesis in cell cultures of carrot.

Rejuvenation: 1. Synonymous with dedifferentiation. 2. Treatment that leads to culture invogoration (such as subculture) or revival (dormancy breaking).

Repelcote: The brand name of a material (dimethyl dichlorosilane) used to coat glassware (soda glass); for tissue cultur containers and for other purposes.

Repression: Altered gene expression resulting in the failure of a specific protein synthesis.

Resistance transfer factor: A plasmid, present in certain types of bacteria such as E. coli, that can impart resistance to antibiotics in animals exposd to them.

Resistivity: The degree of resistance (in ohm-cm or ohm-m) or interference to the flow of an electrical current or the movement of particles from place to place. A measure of water purity; the purer the water, the greater its resistivity. The reciprocal of conductivity.

Rhizogenesis: Root formation and growth; as in root development de novo from callus.

Root culture: Isolated root tips of apical produce in vitro root systems with indeterminate growth habits. These were among the first kinds of plant tissue cultures (White, 1934) and remain important research tools in the study of developmental phenomena; mycorrhizal; symbiotic and plant-parasitic relationships.

Rotary- shaker: A platform shaker with a circular motion used for agitating culture flasks at vaariable speeds.

Rotating culture or rotator: A wheel-like device for slowly (ca. 1 rpm) rotating and gently agitating cultures, usually in a vertical plane.

Scion or don: The portion of a bud or shoot used for grafting onto another plant (the stock).

Screen house: A metal or plastic mesh enclosure similar in function to a greenhouse but with more exposure to the elements.

Scutellum: That part of the cotyledon in Gramineae (grasses) that absorbs food from the endosperm at germination. An explant source for plant tissue culture.

Selection culture: Utilizes difference(s) in environmental conditions or more usually in culture medium composition, such that preferred variant cells or cell lines (presumptive or putative mutants) are favored over other variants or the wild-type.

Selection pressure: The measure of the effectiveness of natural or experimental selection m altering the genetic composition of a population.

Selection unit: Single cells or small clusters, units of optimum size for isolating and regenerating variants or mutants; I the minimum number of cells effective in the screening process.

Selective advantage: Implies the possess of increased fitness within San individual or a population.

Selective agent: An environmental or chemical agent that imposs a lethal or subleihal stress on growing plants, or portion thereof in culture, enabling selection of resistant or tolerant individuals.

Serial float culture: Sunderland's (1977-1979) technique of floating anthers on liquid mediumand subculturing them to new medium at several day intervalsan anther dehiscence, pollen release and development occur, increasing anther productivity.

Shake culture: An agitated suspension culture. Usually a flask (commonly an Erlenmeyer flask) containing the culture is attached to a horizontal or platform shaker, or agitated with a magnetic stirrer. to provide adequate aeration for cells in the liquid medium.

Shakeror platform shaker: A platform fitted with clips for grasping flasks (usually Erlenmeyer flasks), or with surfaces suitable for attaching flasks, with set or variable speed control. It is desirable to adjust the shaking speed for gentle, even agitation of suspension cultures.

Shoot tip graft or micrograft: The grafting of a very small shoot tip or menstem tip onto a prepared seedling or mcropropagated rootstock in culture. Meristem tip grafting is used for in vitro\irus elimination with Citrus and for other plants as an alternative to grafting in the greenhouse.

Somatic cell embryogenesis: The production of embryos from somatic cells of explams (direct embryogenisis) or by induction on callus formed by explants (indirect embrygenesis). These rv.o processes may not be materially different in results.

Somatic cell variant or embryoid: An organized embryonic structure morphologically similar to a zygotic embryo but initiated from somatic (non-zygotic) cells. These develop into plantlets in vitro through dèvelopmental processes that are similar to those of zygotic embryos.

Somatic hybrid: A cell or plant product of somaic cell fusion; as the result of cell or protoplast fusion and implying genomic integration. The process is somatic hybridization.

Somatic mutation: Mutation occurring in vegetative cells or tissues.

Somatic organogenesis: The production of shoots, roots or other organs on somatic tissues of explants (direct organogenesis) or by induction oncallus formed by explant (indirect organogenesis).

Stationary culture: A non-agitated culture, the antonym is shake culture.

Sterilization: The process of making things sterile through 1. Rendering plants non-reproductive. 2. Killing or excluding microorganisms or their spores with heat, filters, chemicals or other sterilants. Dry heat sterilization is useful for metal instruments and glassware. Foil-wrapped items are subjected to 150°C for a minimum of 3 hours in a hot air oven. Steam sterilization or autoclaving is useful for nutrient media, distilled water, paper products and glassware. Solutions contained in glass flasks. plugged with cotton and capped with foil are subjected to 1.05 kg/cm (121°C) for 10 to 20 minutes (depending on the total volume and how it is distributed). Foil wrapped capped glassware are autoclaved in the same way. Filter sterilization through a *Millipore-type* membrane is useful for thermolabile solutions such as those containing vitamins and urea. Chemical sterilization (most commonly sodium hypochlorite) is useful for plant materials in preparation for excision (surface sterilization) and for working surfaces. Some medium constituents that are thermolabile or insoluble in water may be sterilized through dissolving in organic solvents, such as chloroform or alcohol. They are then dispensed to sterile filter paper for solvent evaporation and the filter paper with the residue is added to or below the sterile medium. Metal instruments are flame sterilized by immersion in 70% ethanol until required, then flaming.

Stock plant: The source plant from which cuttings or explants are made. These are usually maintained carefully in an optimum state for (sometimes prolonged) explant use. Preferably they are certified, pathogen-free plants.

Surface sterilization: The removal of plant surface microflora prior to aseptic excision of explants. Surface sterilization is accomplished by immersion of tissue in one of many sterilants, such as calcium hypochlorite or sodium hypochlorite, hydrogen peroxide, mercuric chloride, silver nitrate or bromine water for an empirically determined period of time. Sometimes a preceding ethanol dip or spray is useful, as is the addition of a surfactant to the sterilant. Agitation or vacuum may improve the efficiency of sterilization. Thorough washing in sterile distilled water usually follows, to remove the sterilant, although in some cases this is not necessary (hydrogen peroxide).

Surfactant: A surface active agent or wetting agent; as is Tween 20 or Tween 80, Teepol, Lissapol F, Alconox, etc. These agents act by lowering the surface tension so are common addenda to solutions used to surface sterilize plant materials prior to aseptic excision of explants.

Suspension culture: Cells and group of cells (aggregates) dispersed in an aerated,usually agitated, liquid culture medium. These are obtained by adding friable callus to the medium. The plant species, explant type and treatment; the composition of the nutrient medium; and many other features determine the size and nature of the cell aggregates, which may appear different during each culture growth phase. These cultures are used to study cell division, differentiation and metabolism, in secondary product synthesis or form the basis for single cell lines, callus cultures, somatic cell embryogenesis and many other culture purposes.

including via temperature variation and nutrient limitation.

Tent: The enclosure or process of including plants, such as exvitro plantlets, under plastic or glass to maintdain elevated relative humidity.

Teratogenic: Refers to an agent which induces or increases the incidence of gross structural abnormalities (teratomata) in an individual or a population; as do x-rays.

Tissue explant: An excised plant potion of tissue used to initiate a culture.

Tracer: A substance that can be followed within a reaction or an organism; as radioactive isotopes and certain dyes.

Transcription or genetic transcription: DNA dependent RNA synthesis. The process by which cellullar RNA molcules arc synthesized as determined by homologous DNA sections where their sequence recurs in complementary form (the template or transcription strand).

Transfer room: A small room, sometimes sterilized internally by a bactericidal lamp (u.v. irradiation) and provided with clean (fltered) air; employd for sterile transfer work.

Transfer: 1. Culture initiation; the first placement of an explant in culture. 2. Subculture; relocation of cultures to fresh nutrient medium. 3. chamber; laminar air flow cabinet or hood (transfer hood) or room (transfer room) in which this is accomplished. 4. act or process of transferring.

Transplant shock: Refers to the stress involved when Stage II or Stage 111 cultures are transplanted to soil (Stae IV); many or all of the regenerated shoots or plantlets may die if suitable care is not taken to acclimatize them gradually to the soil environment, particularly prevention of water stress.

Transplant: 1. To relocate or remove to a new growing place. 2. The cultured tissue or explant, relocated or transferred to a new site (in vitro). 3. Stage IV; the transfer of plantlets or shoots ex vitro from aseptic culture to soil. The process is transplantation.

True to type: Applied to a plant or propagation source this term denotes correct cultivar identification and lack of variation in productivity or performance. Verification is determined visually by an expert or through biochemical, serological or other rneans.

Ubiquitous: Occurs everywhere, as do bacteria in the environment.

Ultrasonic cleaner: A device to include high frequency vibration of materials, removing adhering substances from surfaces by mechanical action. This device is useful for cleaning glassware and for disinfecting plant material.

Ultraviolet light (UV): Radiation with wavelength (100-400 nm) at the violet end of the visible spectrum Generated by mercury vapor lamps, it is sometimes used in tissue culture for us mutaaenic properties or to reduce ambient contaminants in work areas due to its bactericidal properties Exposure is harmful to the eyes and skin and is to be avoided by workers in plant tissue culture facilities.

Unorganized growth: In vitro formation of tissues with few differentiated cell types and lacking recognizable structure; as with many calli.

V/V: May indicate simple proportion; as 3;1 (v/v). May indicate percent volume in volume, as the number of cm (mis) of constituent in 100 cm-' (mis) solution.

Vermiculite: Any of many minerals (commonly altered micas) whose granules can expand greatly, becoming highly absorbent. Used alone or more commonly as a component of potting mix.

Vernalin: An hypothetical hormone-like substance found in plant meristematic regions, produced by vernalization. This substance is apparently graft transmissible, but has not yet been identified Different cold requiring species may form different substances during vernalization.

Victorin: A host specific toxin produced by *Helminthosporium victoriae* utilizable for screening oat cells for resistance to the pathogen.

Virus elimination: Thermotherapy, chemotherapy and meristem or meristem tip culture, used alone or in combination have been employed for the elimination of systemic viruses from plants. Plants must repeatedly test negatively for the virus in question for assurance that it has been eliminated.

Virus-tested or virus-free: A plant that appears healthy and repeatedly tests negatively for the presence of one or more identifiable viruses. Such a plant may then be used as a stock or donor plant (explant source) for propagation purposes, and may be certified as virus tested (certified virus tested). The term virus-free is incorrect in most cases, as such a plant may contain one or more viruses which have not been assayed.

Vitamin B complex: Includes a large group of water soluble vitamins that function as coenzymes; thiamine (B1), riboflavin or vitamin G (B2), niacin or nicotinic acid (B3). pantothenic acid (B5). pyndoxme (B12), biotin or vitamin H, folic acid or vitamin M (BC), inositl, choline and others.

Water potential: Governs the conduction of plant water from regions of high to low water potential (towards increasing solute concentration) and is a measure of the free energy status of water in a system. The total water potential of system is the sum of the osmotic component (dependent on solute concentration), the matric component (dependent on water-building substances and surface forces) and the pressure component (usually turgor pressure).

Wick: A filter paper, chromatography paper or a strip of some other material used to support plant tissue above a liquid medium and to transport the water and nutrients to the tissue.

Wild type: 1. The genotype or phenotype of an organism predominating in the control (standard or wild) population, in its natural element. 2. A specific gene predominant in this population.

Woody plant medium (WPM): A modified MS (1962) medium developed for woody plant species by G. Lloyd and B.McCown (1980). It has less nitrogen (both ammonium and nitrate), sodium, potassium and chloride than does MS (1962) medium. It has been increasingly used for the commercial propagation of ornamental trees and shrubs.

Wound tissue: See callus.

Yeast extract: A complex, undefined. B vitamin-containing addendum to some plant tissue culture media. Yield test: Productivity assessment.

Zcatin riboside (rmv 351.4): A naturally occurring cytokinin sometimes used in plant tissue culture media.

Zephiran: The brand name for a disinfecting agent for plant material containing benzalkonlum chloride. Zichl's stain: carbolfuchsin.